THE DIVER'S UNIVERSE

A GUIDE TO INTERACTING WITH MARINE LIFE

THE DIVER'S UNIVERSE

A GUIDE TO INTERACTING WITH MARINE LIFE

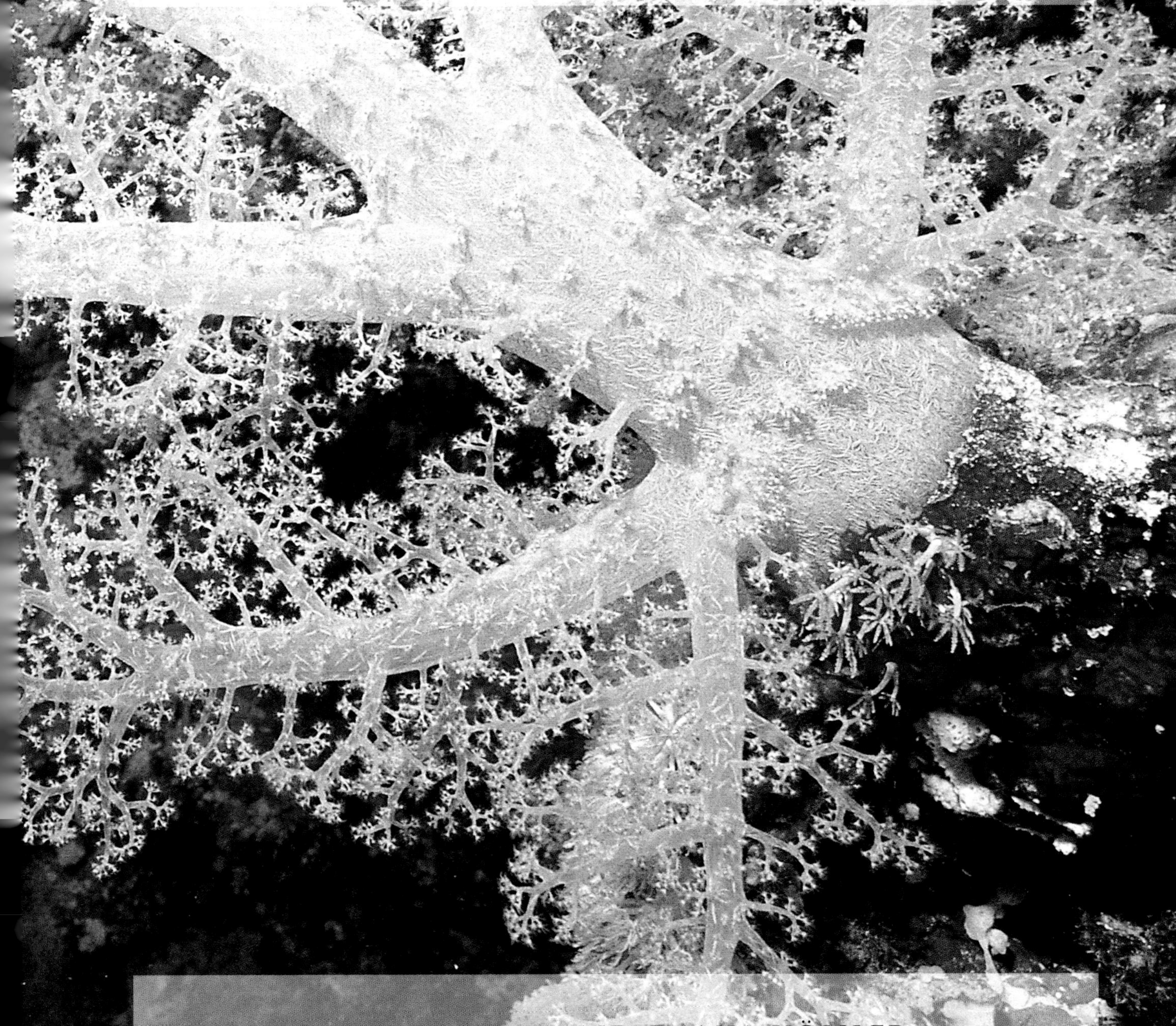

ANNEMARIE AND DANJA KÖHLER

This edition published in 2003 by
New Holland (Publishers) Ltd
London • Cape Town • Sydney • Auckland
www.newhollandpublishers.com
First published in 1997

86 Edgware Rd
London W2 2EA
United Kingdom

14 Aquatic Drive
Frenchs Forest
NSW 2086, Australia

80 McKenzie St
Cape Town 8001
South Africa

218 Lake Rd
Northcote, Auckland
New Zealand

Reproduction by Unifoto (Pty) Ltd
Printed and bound in Singapore by
Tien Wah Press (Pte) Ltd

ISBN 1 85368 642 5 (hb)
ISBN 1 84330 573 9 (pb)

Senior Designer: Lyndall Hamilton
Editor: Anouska Good
Publishing Manager: Mariëlle Renssen
Cartographer: Lyndall Hamilton
Proofreader: Thea Grobbelaar

2 4 6 8 10 9 7 5 3 1

Author's acknowledgements
The author wishes to thank:
• Chris, Suzy and crew on the MV *Telita*, Papua New Guinea;
• Craig de Witt, Scottie and Diane Waring and crew on the MV *Golden Dawn*, Papua New Guinea;
• Rob Van der Loos and crew on the MV *Chertan*, Papua New Guinea;
• Manta Ray Bay Hotel and Yap Divers and, in particular, Leo Tamag, Yap, Micronesia;
• Rob Barrel and crew, on the *Naia*, Fiji;
• Anthony's Key Resort, Roatan, Honduras;
• Gary Adkinson and staff of Walker's Cay, Bahamas;
• Kevin Cock and staff, Paradise Diving, Mauritius;
• The entire crew of *Poseidon's Quest*, Red Sea;
• And all our friends in the Maldives.

Thank you and well done to our superb publishing team, Mariëlle Renssen, Anouska Good and Lyndall Hamilton, who made this a special experience and a special book.

Lastly, to those who gave me the time and quietly supported me from the wings with love, understanding and patience: my husband Helmold and my daughter Amori.

Dedication

To our beloved reefs, glorious symphonies in which Nature plays endless varieties on a basic theme on her most exquisite instruments, the creatures that inhabit them, and…

to Helmold Köhler, infinitely wise and patient husband and father, who had the courage to learn diving at age 67, still joins us on each venture five years later and finances all our expeditions into the blue.

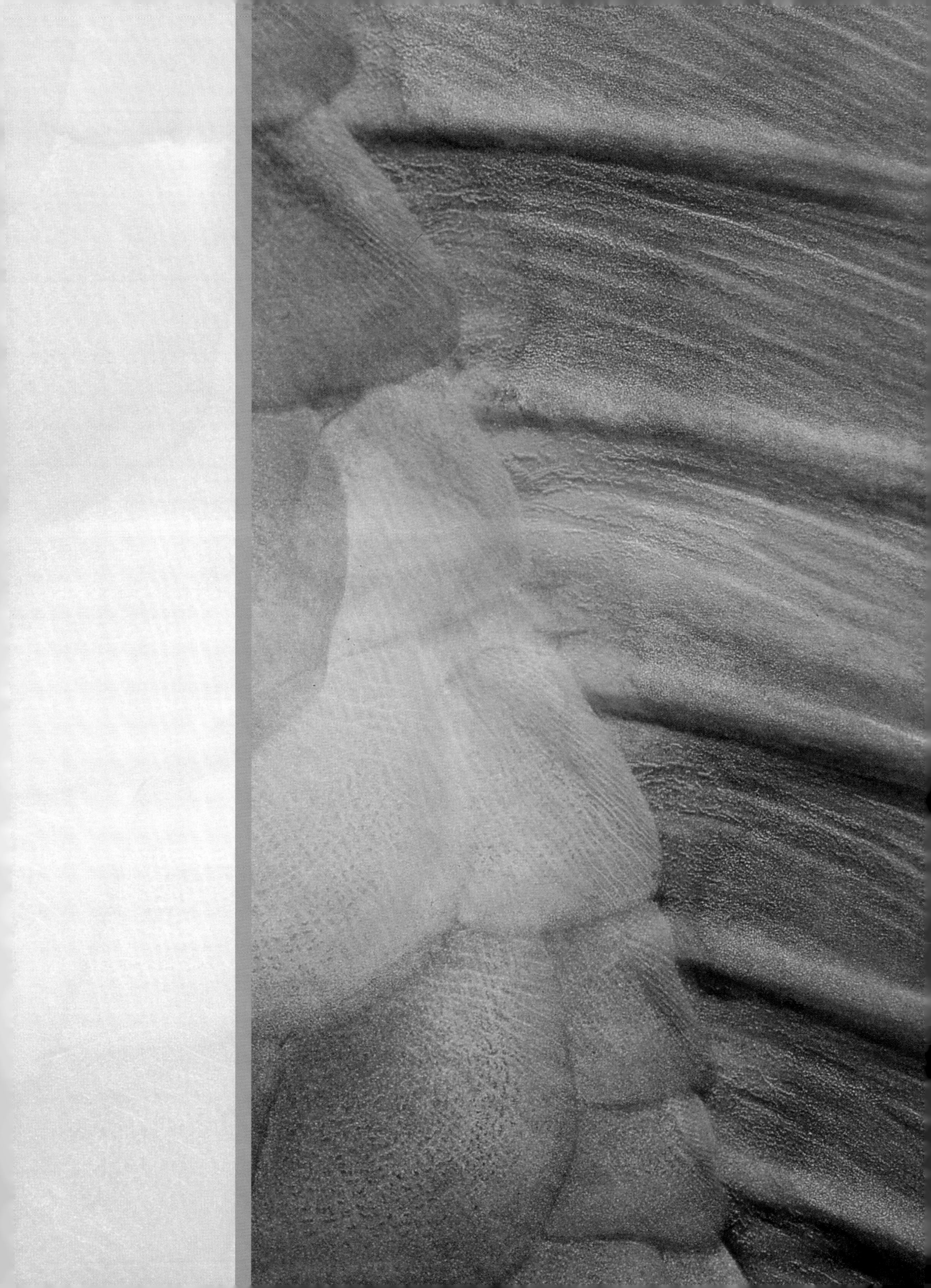

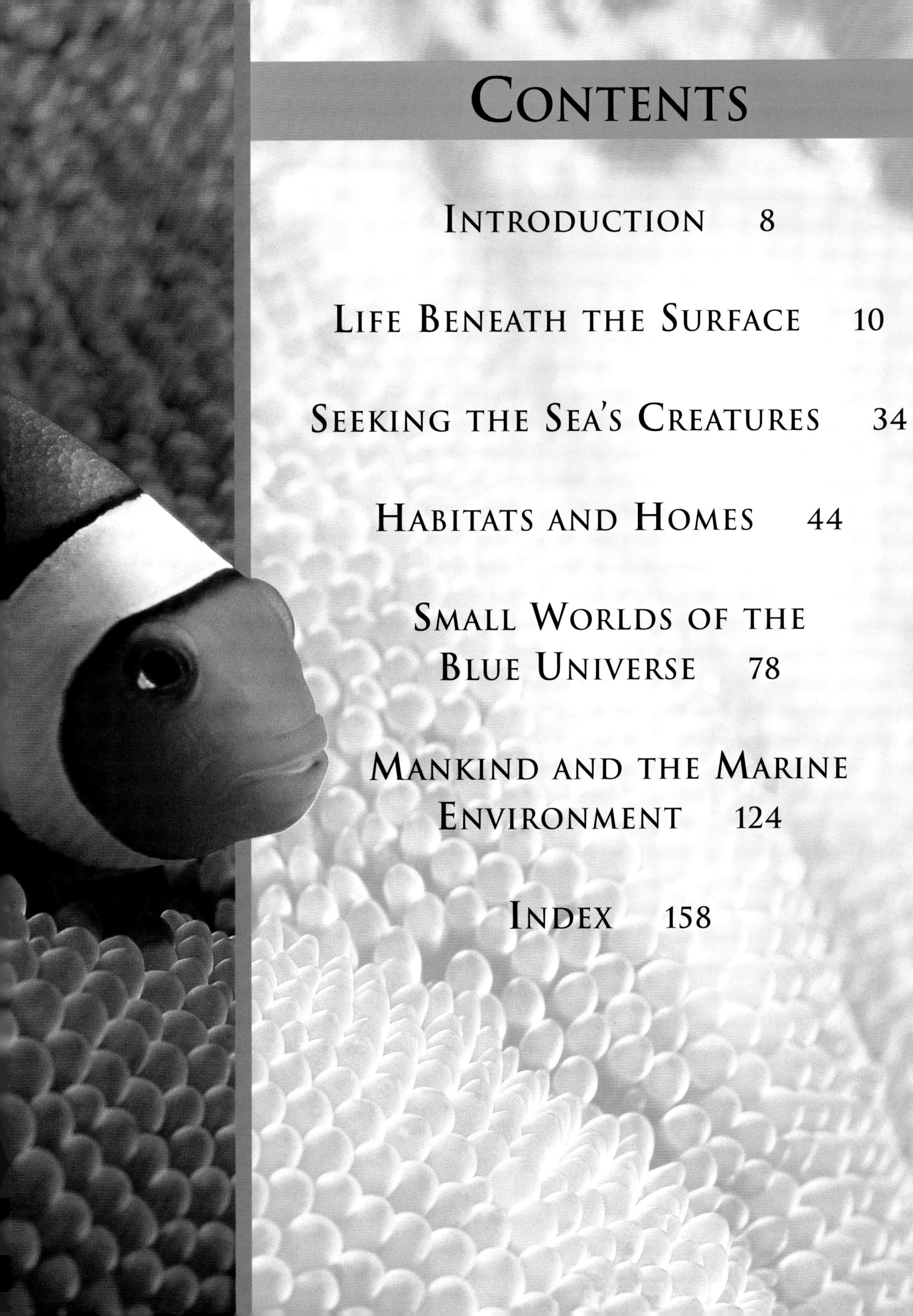

Contents

Introduction 8

Life Beneath the Surface 10

Seeking the Sea's Creatures 34

Habitats and Homes 44

Small Worlds of the Blue Universe 78

Mankind and the Marine Environment 124

Index 158

INTRODUCTION

Unique among the planets of our solar system, our world when seen from space appears blue. Earth, quite simply, is a marine habitat, for oceans cover almost three-quarters of our planet. In fact, almost all the water on earth is ocean – 97 per cent, or over 12.5 million km^3 (three million cubic miles) of it. Thus, the ocean also accounts for 95 per cent of earth's living space. Life began in the sea three and a half billion years ago; on land it has existed for less than half that long.

The oldest natural communities we know, the coral reefs, grow within earth's tropical midriff, a belt that stretches away on either side of the equator between the lines of the Tropic of Cancer and the Tropic of Capricorn. Only in very few places does this girdle flare out beyond its true tropical borders, because of currents that still provide adequate warm water. One of life's largest and most awesome creations, and visible from space, these reefs are the ultimate glorious expression of the phenomenon and tenacity of life. Zoologists and biologists recognize and classify 26 major groups (or 'phyla') of animals, based on body design. Less than half of them are represented on land but all are found on coral reefs in a heady mixture of exquisite beauty and fascinating function.

Yet, sadly, we have come across many experienced and accomplished divers who habitually dive deep but who, for some reason, have never learnt to see the weird and wonderful marine creatures in the very wilderness they dive. Obsessed with salvaging wrecks or encountering the 'big stuff', they blithely pass by the hidden secrets, unaware of their loss. The number of dives logged by a diver is not representative of competence or expertise and it is especially not indicative of his ability to really experience, at close hand, the most intricate mysteries of the reefs.

Few people realize just how totally dependent the largest part of this underwater wilderness is on light for its growth and incredible beauty. Depth or the lack of scuba-diving skills do not necessarily exclude the nondiver from exploring the oceans. The most prolific parts of the reefs are those with maximum sunlight, often the first 10m (33ft). The passing parade of exquisite creatures that thrives here is frequently well within the snorkeller's range. With a mask, snorkel and fins even young children can discover the many wonders of the 'deep'.

After many years of diving and our personal involvement in research for pleasure, we realized that divers have almost no literary resources from which they can learn how and where to find the many smaller, secretive, but equally wondrous creatures of the reefs. That is the purpose of this book. By sharing the knowledge and secrets of photographers and experienced naturalist divers, we hope to make your underwater experiences much more fulfilling. By making you aware of the unmatched designs of marine creatures, the ingenuity employed in their quest for survival and the fascination of procreation on the reefs, we wish to open eyes and minds to the beauty of the ocean and touch the hearts of tide-pool explorers, snorkellers, divers, and underwater photographers alike.

In sharing with you all the magic that binds man to nature in fulfilment of both spiritual and physical nourishment, we hope to contribute towards protecting our planet and winning friends for these most enchanted and mystical places on earth – the tropical reefs.

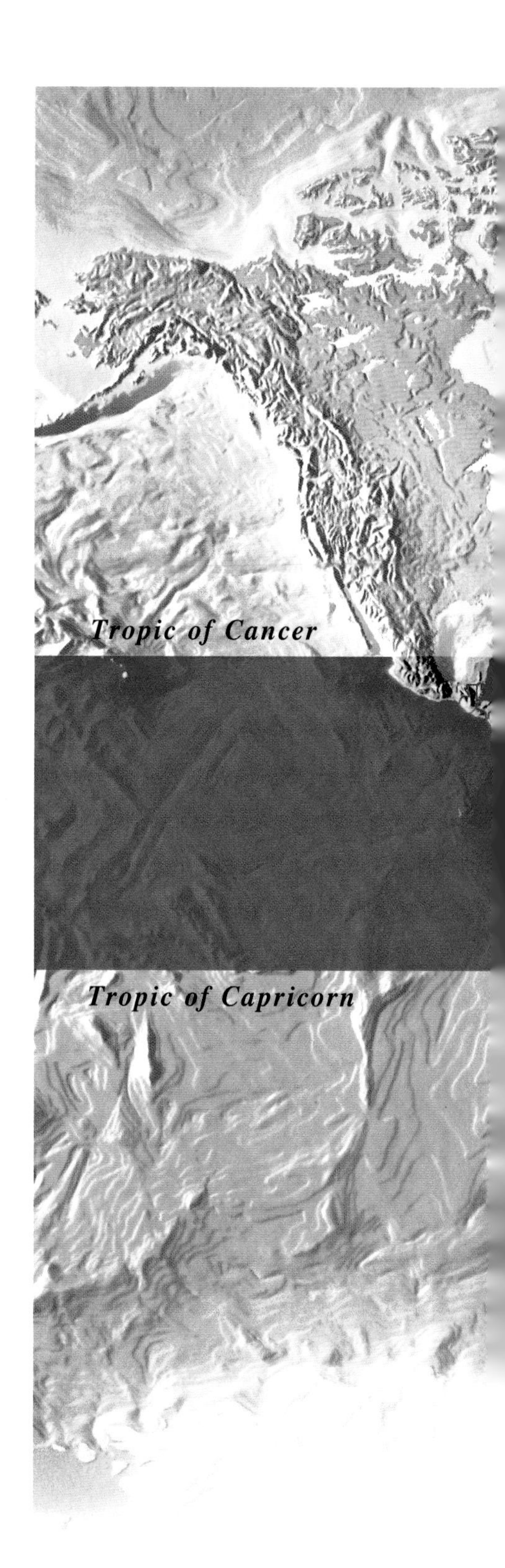

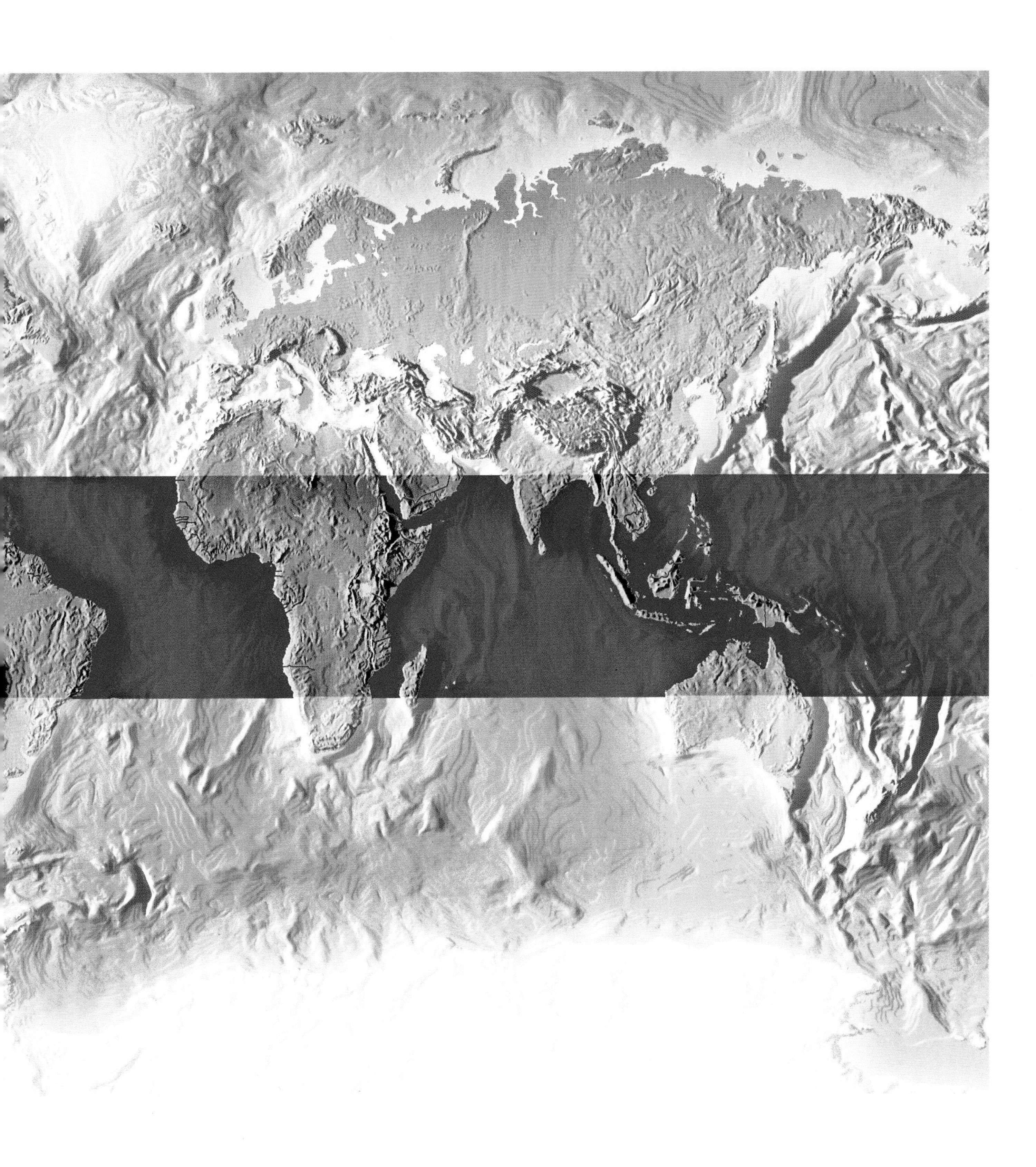

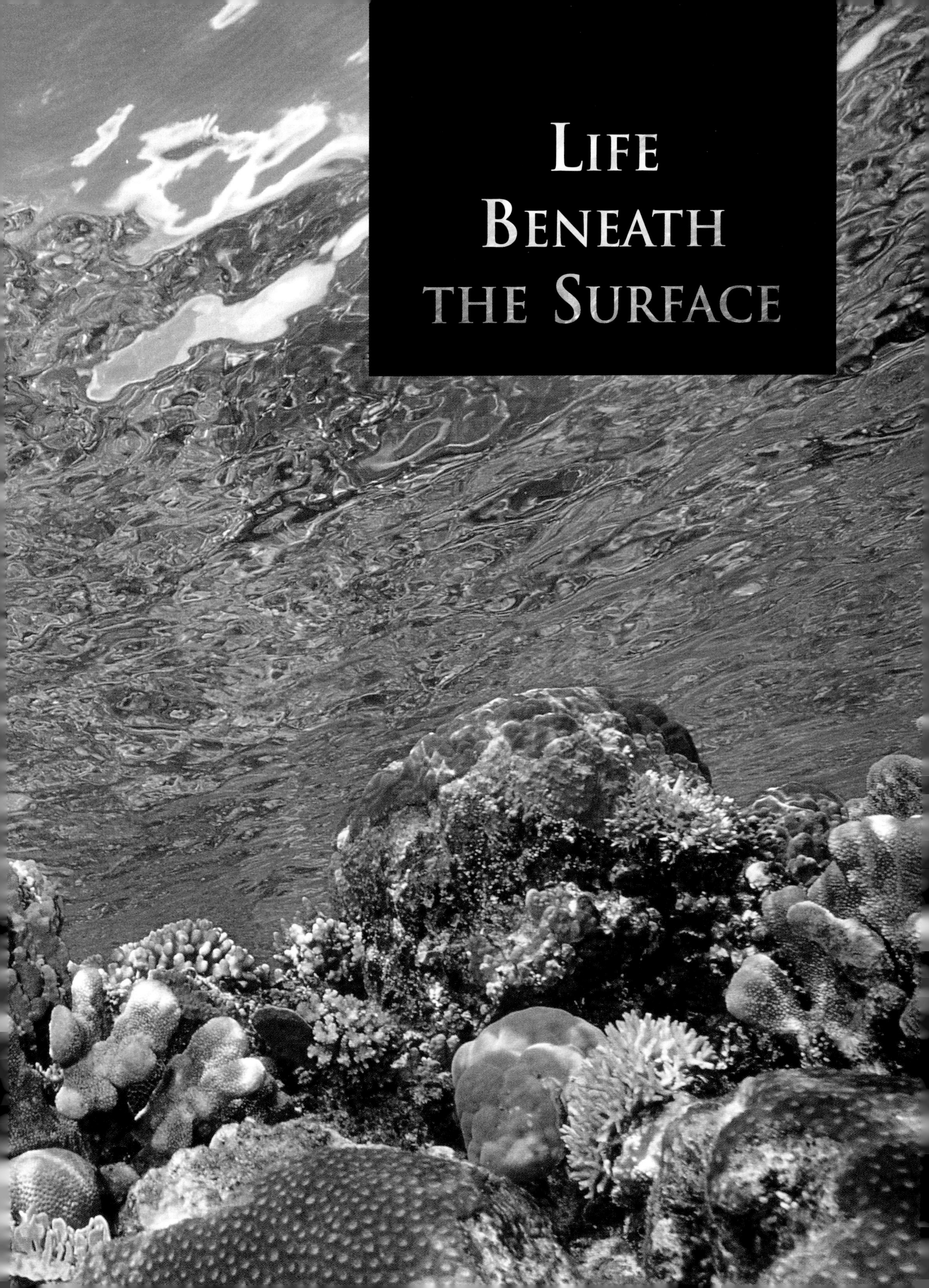
LIFE
BENEATH
THE SURFACE

Cities of the Sea

Previous pages: *Limpid tropical waters and coral gardens create the rich domain of intricately interlinked marine life.*

Below: *The countless coral structures offer safe hides and homes for reef inhabitants, including the redfin butterflyfish* (Chaetodon bifasciatus).

Buffered by warm waters and protected from climatic extremes, the tropical reefs are a dazzling confusion of colour and light, pattern and texture, movement and sway. They are continuously squeezed by tides and swirled by currents, brilliantly lit, sometimes serene and sometimes battered by devastating storms. They tower amidst vast stretches of sandy wastes and they present endlessly varied perspectives every moment of the day.

Paradoxically, the clear, clean waters of the tropics are low in nutrients. Yet a healthy tropical reef teems with abundant life, proving that this drawback has been miraculously overcome by the independent nature of the reef system. The majestic and variable coral reef structures are built because of biological alliances involving a system of give and take that is so subtle and so unexpected that we have only very recently begun to understand it. Scientists call this alliance symbiosis which, in its purest form, is nothing but an agreement to live in peace.

THE PEACE TREATY OF THE REEFS

The secret of success is found in a most amazing partnership, a treaty to which the signatories are an animal and a microscopic plant. The animal partner is related to the sea anemone family. The plant partner is a single-celled marine alga, collectively called zooxanthellae. These partners live as one single entity to form stony coral and together they construct the majestic structural components of what we know as coral reefs.

Of course neither of the partners can conjure organic molecules for life out of nothing. Thus, opting for a sedentary life, they settle in places that are washed by currents; it is on these currents that an adequate supply of imported food from far-off, more nutrient-rich places is transported. There the partnership begins a process that results in sheer natural architectural magic. The carnivorous animal component, rather than having to search for and find food, strains passing plankton from the water with crowns of flexible tentacles which serve as living fishnets. Once the food is digested, the wastes, which contain carbon dioxides and ammonia, become nutrients for the pampered one-celled plants living within their bodies. The plant partner, fed and stuffed with chlorophyll, need not scavenge the environment but can employ the supplied nutrients to harness solar energy through photosynthesis for the production of more food. When sunlight strikes, the plant cells stack into orderly exposed tiers, an arrangement that optimizes the utilization of solar energy and food production. Then, after the partners have exchanged various compounds to their mutual benefit, they produce the porous limestone rock that serves as its own skeleton and as the backbone of the reef. As corals age and die, the new generation settles upon the remains of the previous generation, slowly building forth layer by layer.

Below: *A detail of the hard coral* Diploastrea heliopora.

Below left: *Solitary mushroom coral* (Fungia *sp.*).

Over time, individual colonies grow and fuse, often glued together by various other kinds of coralline algae that produce the equivalent of limestone 'cement'. And thus, day by day, over millennia, tropical underwater cities are built. The skyscrapers, suburban homes and slums of human making find much more magnificent natural duplicates here. Riddled with cracks, nooks, caves and crannies, the structures offer protection to many organisms. Larvae and plants soon settle and begin broadening into a food source for a host of invertebrates and fish.

This 'peace treaty' thus forms the first vital step in creating the coral-reef life and food chain. Once there is adequate food and protection, more takers flock to the reefs. And so, step by step, the web of life is woven; an incredible arrangement of interrelationships, collaboration instead of confrontation, and an ongoing quest for survival between those that eat and those that prefer not to be eaten.

Below and bottom: *Macro shots of soft corals.*

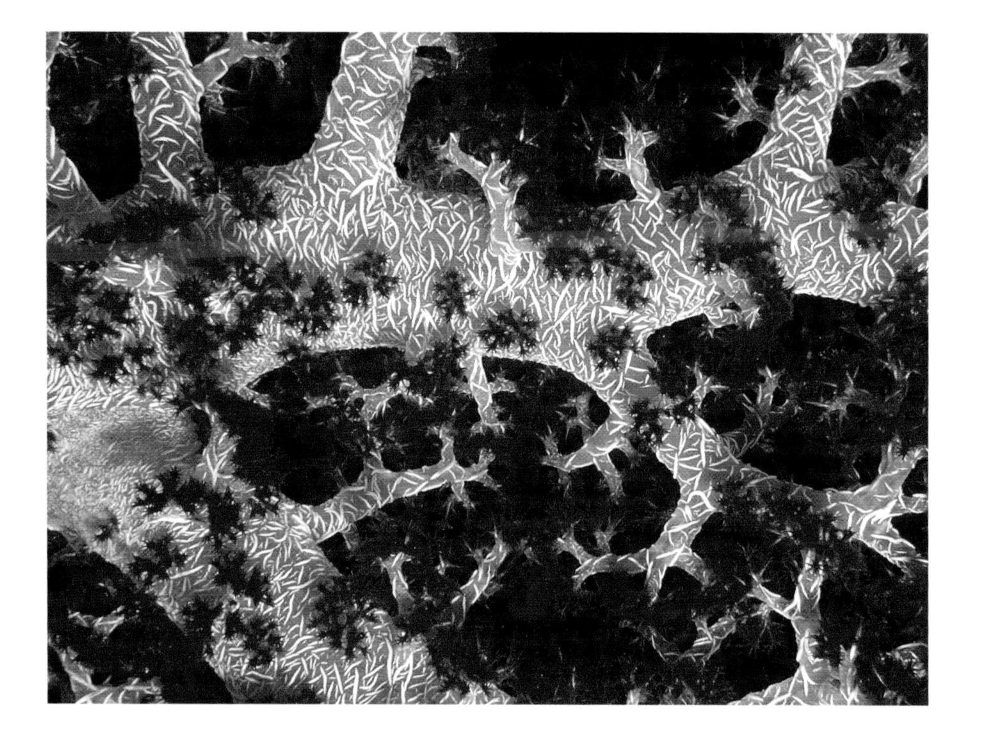

More food in the form of plankton is imported by powerful swirling underwater rivers, currents that begin in immense oceanic areas far from the tropical reefs. This rich living broth containing organic creatures – both animal and plant – that live suspended in water greatly increases the odds for successful life. Larvae, bacteria, yeasts, fungi, algae, protozoans, crustaceans, worms, salps and jellies are all adrift in the oceanic pantry. Although suspended in currents, the planktonic animals are not entirely helpless. Swimming legs, tiny hairs and a multitude of appendages beat to propel them. Some use gas bubbles to stay afloat, some store oil droplets from food to gain buoyancy, some have flaps, folds and fins. Together with the sugar, starch and oxygen-producing plants – phytoplankton – they provide the bulk of nourishing reef energy.

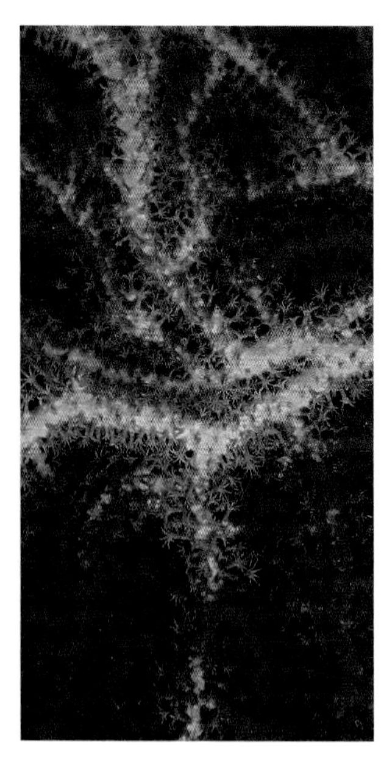

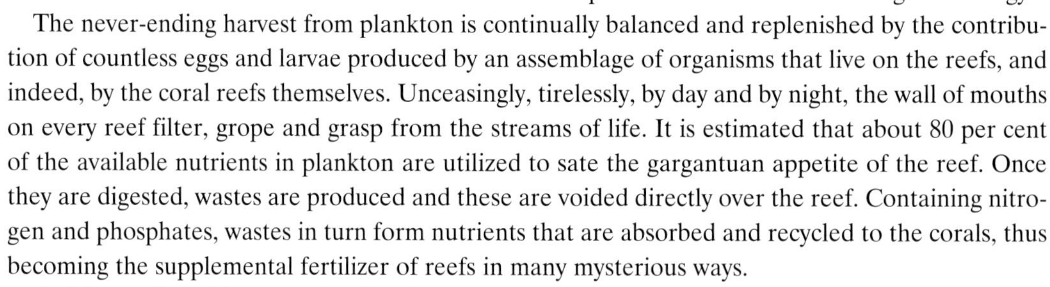

The never-ending harvest from plankton is continually balanced and replenished by the contribution of countless eggs and larvae produced by an assemblage of organisms that live on the reefs, and indeed, by the coral reefs themselves. Unceasingly, tirelessly, by day and by night, the wall of mouths on every reef filter, grope and grasp from the streams of life. It is estimated that about 80 per cent of the available nutrients in plankton are utilized to sate the gargantuan appetite of the reef. Once they are digested, wastes are produced and these are voided directly over the reef. Containing nitrogen and phosphates, wastes in turn form nutrients that are absorbed and recycled to the corals, thus becoming the supplemental fertilizer of reefs in many mysterious ways.

A rich reef is exciting and visually stimulating. For the eye new to it, the colour and diversity can be overwhelming, often leaving novice divers with a helpless feeling of sensory overload and considerably hampering the immediate absorption of detail. It is only after repeated visits that one begins to make sense of the whole. One suddenly begins to see the harmony between patterns and light, and to pick out individual species. For quite a while the darting anthias, expanding out from corals and zipping back at the slightest threat, the gaudy reef fish, the exquisite soft corals or the giant gorgonian fans will be the greatest attraction and leave the most enduring impression.

But there is a hidden wealth of fascinating life in every reef tapestry, ready to be seen and explored. All that is required is to learn to truly open our eyes and behold.

FISH OF THE REEF

A reef can only thrive in almost constant sunlight. In the shallows it is brilliant and lush; with increasing depth it turns ever more inky-blue and bare. This creates many finely divided layers, each of which is suited to different reef organisms. While it is our intention to concentrate on creatures that are less known or less evident on the reef, we must always see them in the context of this entire system. At first, the fish steal the thunder most of the time and it is perhaps they who best convey the health and richness of a coral reef.

Of an estimated 12,000 marine fish species, about 7000 are found on coral reefs. Naturally, it is therefore outside the scope of this book to discuss all these fascinating inhabitants or, indeed, the magnificent coral fortifications themselves. Faced with choices between creatures, I have tended towards those that are within reasonable reach of a wide range of divers and that still have unusual traits. I have also tried to concentrate on creatures that demonstrate concealment strategies well, because once understood, they will help most in the development and growth of a diver's awareness and vision.

I will therefore touch only briefly upon the 10 main reef fish families. Throughout the ages, these particular species have divided the utilization of reef resources into distinct parts. Contrary to the generally accepted theory of 'survival of the fittest', their lifestyles and feeding habits often involve great tolerance, contributing to a surprisingly peaceful coexistence and even active collaboration at times. It is this ability to cooperate and diversify that is the strength of the reef community.

Most fish do not wander from one reef to the next. They are either territorial or at least home ranging. These territories are chosen and shared or defended selectively. Fish may aggregate in groups of their own families or may be loners, and any might rob the others of their food supply or contribute to it. In general, though, the diversity of both food and species places them in different nonconflicting slots on the reef. Even species that are not strictly territorial confine themselves to a particular section of the reef where the available feeding grounds determine the size of their range.

Above: *A spotted hawkfish* (Cirrhitichthys aprinus) *peeps from its coral perch.*

DAMSELFISH (*Pomacentridae*) are the most prolific and indeed the most evident species on the reef. Their excited grunting and clicking chorus can be heard clearly throughout every dive. Typically found in loose congregations, they are active during the day and are mainly plankton feeders. The small sergeant-majors, brilliant lemon damsels and iridescent chromis are present on almost every reef. Some damselfish zealously guard small plots of algae-rich turf, home territories where they have control over their food supply and nesting facilities. Irrespective of size, they will cheekily and with fearless vigour chase away all intruders. This strong territorial urge makes it easy to watch their courtship behaviour. After spawning, often with several females, the male damselfish guards and cares for the nest of eggs until they hatch. At this time males are very aggressive, even towards much larger egg-eating thieves such as butterflyfish or wrasse. While the eggs of most damsels drift away as larvae, the brood of the spiny chromis (*Alcanthocromis polyacantis*), once hatched, gather into a miniature school which the parents continue to feed and protect until their offspring are able to look after themselves.

▼

▲

WRASSE (*Labridae*) live in and exploit all reef environments. Although many feed on small invertebrates on the reef surface and delicacies in mid-water, diets vary widely between different wrasse species. They utilize almost all the different food sources. The small opportunistic chequered wrasse often follow sand-digging fish or sand-kicking divers to profit from the stirred-up food supply. Wrasse sport such varied colours and body forms that it is almost impossible to believe some are related!

Sex change is a survival phenomenon present in many species. A number of wrasse begin reef life as 'primary' or original males. Their sex organs differ from those of 'secondary' males, which can develop from females as and when the need arises. The lunar cycle plays a major role in triggering their spawning. Wrasse normally court and spawn around dusk, sometimes in pairs but often in small groups in which the most colourful males seem to dominate. Under protection of twilight, the spawn rises to the surface and floats away.

BUTTERFLYFISH (*Chaetodontidae*) are best known for their exquisite colours, intricate patterns and elegant ways. These graceful fish bond faithfully for life and are almost always found in pairs. Many butterflyfish feed partly or exclusively on live corals, but some may be seen nibbling at stinging organisms such as sea jellies and anemones. They tend to be shy and patience is required to observe them closely, as their flattened bodies are perfectly suited to slip easily into cracks and crevasses or between the protective branches of corals. Their eyes are often hidden by black bands, and many sport prominent false eye-spots. The picturesque bannerfish and the dainty Moorish idols with their long trailing dorsal fins are members of this family. ▶

▲

ANGELFISH (*Pomacanthidae*) are closely related to butterflyfish and some may be confused with them. The most noticeable distinguishing characteristic is a powerful spine on the lower edge of the gill cover. Beautiful and stately, angelfish are somewhat more solidly built than their butterflyfish cousins. They often have elaborate stripes or vibrantly coloured features which are used both in territorial display and for quick recognition of mates. Colour patterns are thus of particular importance to those species that sport them. Angelfish feed on invertebrates and sponges, especially when damage to these organisms' tougher outer skin reveals the more tender inner flesh.

When agitated, they are capable of making a rather loud drumming noise, a penetrating sound which often catches unwary, intruding divers completely by surprise.

The juveniles of several angelfish sport colours and patterns so vastly different from their parents that they could quite easily be mistaken for totally different species.

THE GROUPER AND BASSLET (*Serranidae*) family accommodates some of the largest as well as daintiest fish on the reef. Unbelievably, the majestic Queensland grouper, which grows to a length of 270cm (105in) and can weigh over 400kg (880lb), is kin to the tiny quivering anthias, better known as fairy basslets, or goldies. Most fish in this family are capable of the female-to-male sex change. Anthias schools consist of several small harems jealously managed by a spectacular dominant male. If such female harems become larger, more males will guard them. Anthias are the most colourful members of the family, varying from vibrant orange in the Red Sea to bicolour yellow and pink in the Indian Ocean to golden-pink, orange or lush magenta on the reefs off the Solomons and New Guinea.

◀

▶

CARDINALFISH (*Pogonidae*) are mostly nocturnal, though some seem quite happy in sunlight and even seek it deliberately. These small, slow-swimming fish abound on reefs, always in groups. During the day they rest: mostly in rocky areas, among multi-branched corals, in caves, and under overhangs. At night they emerge to feed on shrimps and crabs, and are then the most abundant species seen. This family makes use of a curious reproductive method – mouth-brooding. Watch for the obviously bulging cheeks of males who incubate the fertilized egg-mass. Every so often, the eggs can be clearly seen when the male 'churns', or juggles and repositions them, for equal egg development. Until the eggs hatch the father will forego all food. Juveniles often shelter among the spines of sea urchins. The cardinalfish most often encountered during the day is the yellow *Apogon cyanosoma*, or goldstripe cardinalfish.

PARROTFISH (*Scaridae*) are very closely related to wrasse and sometimes, when displaying similar colours, may be confused with them. However, parrotfish have gained their common name from characteristic fused teeth that resemble parrot beaks. With these they bite, break off or scrape corals to feed on the algae – their bite marks can be seen on many hard corals. But, although they consume copious amounts of coral rock, they always move on to others after a few bites, naturally pruning but never destroying the reef. The coral is ground to a fine powder by specialized teeth in the throat and later voided with faeces. The resultant fine, sifting sand-rain contributes much to the vast expanses of sand. Individuals from this family also make use of the female-to-male sex change in times of need and loss. ▶

SURGEONFISH (*Acanthuridae*) are frequently seen in loose schools or congregations which graze on vegetation and the filamentous algal mat that covers the reef. At times they may gather for mass migrations. Their common name is derived from the very sharp, scalpel-like structures at each side of the tail base, which in some species are kept sheathed and in others may always remain erect. Divers should refrain from handling these fish as the cuts caused by them can be serious and painful. Unicornfish also belong to this family. Surgeonfish often enter holes and crevices, easing their tails in first.

▼

Following pages: *Exquisitely transparent glassy sweepers* (Parapriacanthus guentheri) *have strong schooling instincts.*

BLENNIES (*Blenniidae*) boast a large family and can best be described in one word: cute. They are small territorial bottom-dwellers, either scaleless or with very small scales, and may easily be confused with gobies. Mostly solitary souls, blennie pairs sometimes occupy adjacent holes. Less frequently, small groups can be found living close together. Lovingly called 'little beauties' by divers, they peek cheekily from holes in corals and tube sponges, from empty shells or from the evacuated tubes of marine worms. They often sport tiny tufts on their comical, toad-like heads and continuously roll their large eyes, observing the reef. Blennies are truly the nimble sprites of the sea but, although very active, few swim around for long periods. They prefer short journeys, mostly undertaken in small 'darts' in search of food.

The blennie male is quite the man-about-town and spawns with many females. Eggs are deposited and hidden in crevices and under rocks, where the male will guard the nest until hatching takes place. The midas blennie (*Ecsenius midas*) sometimes swims upwards to mingle with groups of anthias of similar colour, seeking protection in the school. But it is easily recognized by its less graceful movements for, unlike anthias, blennies have only a residual swimming bladder and, as a result, must move their tail continuously for propulsion and stability.

▲

THE GOBIES (*Gobiidae*), in terms of sheer variety, are the largest known family of fish, with more than 1600 documented species. Their identification is truly a taxonomic nightmare for, although they are the most abundant fish, their very small size and timid, secretive habits make them difficult to see and study. Most are smaller than a human little finger; the smallest known vertebrate in the world, *Trimmaton namis*, belongs to this family and measures just 8mm (0.3in) long!

Very close inspection of rock, sponge and gorgonian coral surfaces is necessary to detect the tiny fish. Its transparent body is adorned with microscopic patterns that perfectly match its home turf. Found in both very shallow and very deep waters, one cannot but appreciate the tremendous diversity of this group. The deep-water species have the most exquisite colours. Other interesting and spectacular gobies live in sediment or sand and either collect rubble to build burrows or simply bury themselves. The partner gobies form lifelong bonds with almost-blind burrowing shrimps of the *Alpheidae* family. The fish serves as a sentinel, keeping constant close contact with the visually impaired shrimp, using tail movements to warn or reassure it. In return, it obtains a safe, ready-made and regularly maintained burrow for retreat.

Gobies are distinguished from blennies by the difference in dorsal fins. In gobies this fin is always divided into two sections, in blennies it is continuous. Gobies have more slender, fishlike faces and live camouflaged on reef organisms, while blennies have broader, fleshier, almost froglike faces and in all likelihood peep out of holes. Gobies are also more readily found in pairs for they are conservatively monogamous. The best known and most frequently encountered species is the skittish fire goby (*Nemateleotris magnifica*) with its long, white dorsal flags and typical hover-and-dart dance routine above its burrow.

. . . And They Multiplied

Reproduction on the reef is an astonishingly rich, primal and varied chorus of life. In the quest for self-perpetuation, extraordinary strategies are employed to keep the legacy of the next generation safe.

In the seemingly chaotic organization of nature, many species are required simply to abandon their destiny to the games of chance. The biggest gamble is external fertilization and it takes place against overwhelming odds.

Fixed organisms and slow movers like clams, corals, urchins and sea cucumbers must not only synchronize the simultaneous emission of untold numbers of eggs and sperm, but must do so while keeping their own male and female gametes separate to avoid self-fertilization. They do so in a spectacular orgy of gigantic proportions. The time is mysteriously signalled once a year by a full moon that coincides with a rise in temperature, low tidal fluctuation and minimal water motion. Somehow the sessile organisms agree, in complete unison, exactly when to ripen and ready their sexual products. A week later, during strong tidal currents, a simultaneous mass spawning over hundreds of kilometres of coral reef begins. For a week of nights, the collective release of unbelievable amounts of eggs and sperm maximizes the chances of fertilization. In the dark, egg-eating predators cannot see spawn carried on the currents. By day, they are overwhelmed by an opulence of seasonal food that they simply cannot consume entirely. The rich, fatty spawn-slick can be seen on the sea's surface, stretching for kilometres in the currents. Tightly packed, these specks may individually be smaller than sand grains, but they travel across the vast open oceans and seas on a grand voyage destined to create future reefs. By the time they finally settle, many of their parents may already be dead.

Fish that launch the new generation by scattering their eggs in the open waters all do so on an outgoing tide and preferably in the safety of dusk. Almost all of these fish display a behaviour called 'looping'. It entails a process of courtship that displays much fast swimming, chasing and posturing, a rapid rise towards the surface, culminating in an explosive release of gametes at the apex and a rapid descent to the reef. On the float, hundreds of eggs die or are eaten for each one that lives.

Bottom-nesting species all have safe-keeping instincts. They clean and prepare suitable surfaces before depositing their eggs in safe, deeply hidden or fiercely guarded nests. Their courtship behaviour entails aggressive chasing and elaborate fin display to ward off intruders and among them parental care, either by one or both partners, is the norm.

Above: *Thousands of jellyfish follow the path of the sun at Jellyfish Lake, Palau.*

Below: *A sea cucumber* (Holothuria fuscopunctata) *releases its spawn.*

Some creatures attach tough, leathery egg-cases to the substratum and abandon them. Cuttlefish and squid 'plant' delicate, jellied egg-sacs on protected cave ceilings or at the base of branching corals and grasses, but then abandon them. The octopus, however, lovingly cares for her brood until they hatch. Their destiny complete, many species from this family will die.

Female crustaceans carry eggs beneath the abdomen until they hatch, while sea horse males willingly submit to a state of 'pregnancy' by hatching the females' eggs in special belly pouches.

Cross-fertilization is very important to all organisms. The strength of a species depends on making the best choices possible from the available gene pool. Thus there are many mechanisms which help shuffle, mix and disperse the different, individual genes and as many precautions to prevent self-fertilization. Most organisms avoid this simply by having different individuals carry the opposite sexual products. In those that carry both male and female organs, the hermaphrodites, several anatomical or biochemical blocks may be present. Otherwise, such species may either recognize their own gametes and refuse to react with them, or have sex products that mature at different times. Cross-fertilization, of course, also allows small physical variations through future generations and this, ultimately, is the raw material of evolution. Each successive variant adapts and becomes more suited to its environment if changes have taken place over time within that environment.

Only the higher life forms utilize internal fertilization and the process of birth. The phenomenon of cannibalism is perhaps one of the most amazing and bizarre aspects; even before they are born, the earliest developing pups of some sharks have already fed on the eggs and siblings that were fertilized later. But to us, mammals like whales and dolphins are more endearing. They, much like humankind, have sex for fun, engage in prolonged and pleasurable foreplay and mate belly to belly. They give live – and often assisted – birth, cuddle and embrace, care for, discipline and teach their young.

Below: *The bottlenose dolphin* (Tursiops truncatus) *is one of the ocean's most endearing creatures.*

Learning to see Order within Confusion

If life were built like a chain, the breaking of only one link could destroy it and complete species would be wiped out. But life is structured like a web, all creatures are flexibly interdependent and the occasional tear in the web is neither dramatically noticeable nor devastating. The quest for survival of any one species depends on its ability to function flexibly within the web. A coral reef is just such an adaptable web of interrelationships; it is self-maintaining and, left to nature, the dynamics of this environment remain surprisingly constant.

For life to exist on the reef, the ability to feed, reproduce and avoid predation is essential. Each creature has developed a distinct lifestyle to allow it to exercise these basic needs in its own unique way. To be successful explorers, it is crucial that we understand this.

The very gaudiness of the reef is the cause of initial diver confusion. But what we perceive as exquisite aesthetics has a deeper and more definite purpose and serves very practical needs – colours, patterns and shapes have evolved in direct correlation with specific life functions and requirements. All have to do with concealment, confusion or advertizement. This mastery of design is a formidable match for the many challenges that are thrust at living organisms by a continually changing, fluctuating environment.

We are tempted to conclude that there is no way a reef creature can remain inconspicuous yet still browse, filter plankton, stalk prey, guard a territory or reproduce. If we only considered the fish, this would be largely correct. However, many other creatures of the reef manage to do just that! Unique problems exist for every layer of life in the water; for each an equally unique solution has been found.

Even if the strength or size of a marine animal teases us into thinking that concealment is not necessary, a predator must stalk and catch its prey. It is the drab colour of the shark, which itself cannot see colour, that allows it to appear mysteriously out of nowhere. The blue-black upper surface and the silvery-white belly of large open-water fish is a commonly used strategy, for in this way they blend with the dark ocean when seen from above and with the bright silver surface when seen from below. Similarly, the dazzling play of light reflected on the silver scales of the schooling surface- and mid-water fish enables a vanishing act in bright tropical sunshine.

Above: *A profusely populated reef is visually overwhelming. Divers must learn to concentrate on individual species.*

Top: *Fin detail of a parrot-fish* (Scaridae *sp.*).

Above: *A detail of the delicate pattern of a peacock flounder* (Bothus mancus).

Above right and right: *The adult emperor angelfish* (Pomacanthus imperator) *(above right) differs remarkably from the juvenile (right).*

On the vivid tapestry created by coral gardens, vibrant reef fish possess pigmented skin and seem to loudly defy the theory of concealment. But they are meant to be seen and almost all of these fish can change the intensity of their colours with ease. Reef fish are able to perceive colour as well as we can, and among the crowded communities that inhabit this environment, these advertizing patterns play an important part in complex social behaviour. They serve as 'team' colours, according to their kind, sex and age, as instant recognition is an important energy- and time-saving factor in mating and reproduction. Males sport the gaudiest colours and will heighten them to proclaim that they are the sexiest and strongest fellows around. Among these fish, juvenile patterns are frequently markedly different from those of the adults.

Yet, even these vibrant inhabitants use clever concealment tricks. Although they are dazzling when seen close-up, they surprisingly 'disappear' when seen from afar as their colours blend with those of the reef. The exotic stripes and patterns help emphasize body shape and desirability from nearby, but also hide weak or vital areas of the body. From afar, they serve to break up outlines. Dark eye-bands and prominent false eye-spots are used to confuse predators and to direct attack away from the real eye. Some fish have fins or supplementary decorations that either help lure prey or deter predators. Others sport flamboyant fins that mimic or blend with spectacular marine growths. A small group can pucker and point their skin in imitation of seaweeds, and change colour almost instantly to blend imperceptibly with their surroundings.

Surrealist or geometric designs are abundantly utilized on the reefs. Stripes, spirals, blotches, dots, stars and squiggles are not only present but executed in the wildest combinations. Each single member of a species, while sporting family colours, is unique – similar to human fingerprints.

How these colours and patterns are employed to keep marine creatures from being seen and becoming a snack is something we must grow to understand if we want to learn where to find them.

THE CLOAK OF DECEPTION

Unquestionably, the greatest form of protection is to be an almost indistinguishable and integral part of the reef. So it is here that we encounter perhaps the most stunning, most varied and most highly developed strategy for survival: deception.

Deception by colour, pattern or mimicry is a widely used and extremely successful strategy. Disguises not only conceal the defenceless and timid, they also allow aggressive species to lurk undetected. Thus, it is vital to understand camouflage. Without it the underwater explorer will fail to find the majority of small marine creatures.

With deception by camouflage, an animal's aim is to blend inconspicuously with its background. While this background could be the reef, it is equally often another living organism. With deception by mimicry, animals adopt the guise of other creatures or organisms that have a valid reason to be avoided – thus either appearing dangerous, poisonous or inedible, or otherwise simply innocent and harmless. This strategy is often employed to stalk unwary prey.

Stonefish and their scorpionfish relatives are masters of disguise. They all use elaborate leafy and lacy growths to melt into the reef surrounds and unobtrusively lie in ambush. One of the rarest, Merlet's scorpionfish (*Rhinopias aphanes*), so perfectly blends with its surroundings that its rarity probably has more to do with divers' inability to spot it. Its colours change to match dominant environmental hues while a plethora of frilly fins and transparent window patterns allow the fish to virtually 'disappear'. We have learnt to use a rather dubious ploy to find it. We look for feather stars in unlikely places, like reef floors and caves. Merlet's scorpionfish seems to imitate the less brightly coloured feather star species which abound where this fish prefers to live. A poor swimmer, it tends to 'walk' along the bottom on specially adapted pectoral fins resembling fingers.

The small sailfin leaf fish (*Taenianotus triacanthus*), also a member of the scorpionfish family, utilizes the movement and colour of surrounding seaweeds as a model, matching and even miming algae patches or surface texture and damage. Enhancing its bluff, it mimes the surge-induced to-and-fro swaying movement of the leaves. It is so confident of its camouflage skills that it sits quite openly in favoured places, swaying gently and occasionally toppling – even when there is no turbulence at all! A bad swimmer, it remains in the same spot for life, periodically moulting as it grows.

Above: *A yellow leaf fish.*

Top: *The elegant squat lobster* (Allogalathea elegans) *assumes the colours of its crinoid host.*

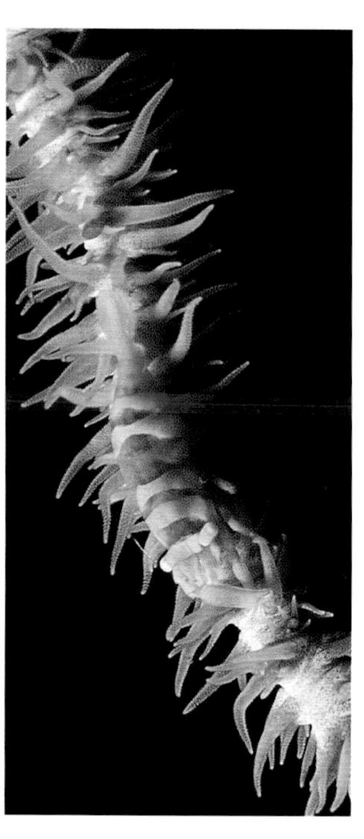

Above: *Using a matching pattern and texture, the wire coral shrimp* (Dasycaris zanzibarica) *blends successfully with its host.*

Above right: *Without artificial light, this tassled scorpionfish* (Scorpaenopsis oxycephalus) *almost defies detection.*

The decoy scorpionfish (*Iracundus signifer*), an ambush predator of the Indo-Pacific, has a stunning fish-shaped dorsal fin which is decorated in great detail with a clearly visible eye-spot, mouth and fin outline. This exotic fin lures carnivorous fish in search of prey straight into the scorpionfish's jaws, because the hunter mistakenly perceives the lure to be potential prey.

Flounders, soles, crocodilefish and wobbegong sharks are all shaped to lie flat and have adapted to grow delicately fringed tufts and intricately patterned marks to merge with the sandy reef floor. The crocodilefish even has a lacy pattern across the top of its eyes, which appears to be an eyelid but is actually an intrinsic, movable part of the eyeball.

The most prolific mimics on the reef are fish. Favoured models are often poisonous or venomous, but there are exceptions to the rule. A small, harmless wrasse (*Labroides dimidiatus*), which provides a cleaning and sanitizing service to other reef inhabitants, is cunningly mimicked by the sabretooth blennie (*Plagiotremus tapeinosoma*). This duplicitous little blennie not only adopts the colours of the true cleaner wrasse, but also copies its stylized jerky dance close to cleaner stations. To the casual eye, the deception is quite believable. The bogus cleaner is often successful for it is initially accepted by unsuspecting clients, at least until they get bitten!

Because of their unique venomous fangs, sabretooth blennies are also successfully mimicked, almost always by other harmless blennies. These already have the correct body shape from their common ancestral heritage and merely need to assume the 'dangerous' colours. Even entirely different fish species have attempted this disguise with relative success. In modelling the colours and shape of the feared fanged blennies, they all obtain a certain measure of immunity from being eaten.

Below: *In crinoids, the ornate ghost pipefish* (Solenostomus paradoxus) *simulates the feathery features of its host.*

The wily trumpetfish (*Aulostomus chinensis*) often swims tightly aligned with the back or flanks of another fish. It deliberately chooses harmless plankton or crustacean feeders, like pufferfish or wrasse, normally ignored by the population at large. From this 'piggy-back' position, the trumpetfish places itself perfectly for a sudden treacherous dart. Its slim, fluted body adopts an inconspicuous colour to emphasize the markings and shape of the host, which can approach the intended prey without causing alarm. In other instances, a trumpetfish may partially adopt the colour of intended prey. Its head alone, for example, may turn yellow to match the yellow fish it is stalking, while the rest of its body remains a blending, almost translucent grey. The unwary prey accepts the colour blotch as one of its own, thus awarding the suddenly pouncing trumpetfish a reasonable measure of success.

Fish are able adapters; when an opportune decoy can be found, they will use it. Trumpetfish and flutemouthed cornetfish (*Fistularia commersonii*) even employ divers with impudence! Along reef walls, the diver's arms, legs or air bubbles offer a perfect opportunity for temporary camouflage from which to dart at fairy basslets (*Anthiinae*). I was also unknowingly used once as a decoy for an extremely well-mannered cuttlefish which 'crept' up behind me and hovered while I was filming a partner goby, devouring it promptly as soon as I was finished.

With this kind of intelligence, the cuttlefish (*Sepiidae*) is probably the most masterful confidence trickster on the reef. It elegantly bridges the distinction between camouflage and mimicry. It has a well-developed brain, sophisticated eyes and a veritable bag of instant tricks and tactics to conceal itself successfully. If you should come across a cuttlefish on the reef, do not take your eyes off it, even for a brief moment! This conjurer's vanishing act is truly astounding.

ARMAMENTS AND DEFENCE IN THE MARINE WORLD

Under natural circumstances, attack for any purpose other than predation, and perhaps dominance in mating, very rarely occurs. Animals on the reef are neither as aggressive nor as culturally obsessed with victory as Homo sapiens. Even when reef creatures protect their nests, they will feign ferocity, but will submit with fatalistic resignation if they are overpowered. Their own survival is necessary to preserve their potential for future reproduction. All armaments are therefore utilized infrequently and then almost exclusively in defence.

Coral reefs are kingdoms of predation and this has led to miraculous inventions for survival. Besides camouflage and cooperation, which evolved largely to avoid energy-sapping conflict, there is a proliferation of exotic and effective weapons spanning some of the most lethal poisons and toxins known to man, as well as an array of unbelievable blades, darts, spears, bristles, and spikes. Yet we should remember that these weapons are designed to cope with threats in the natural underwater environment – man was not intended to be one of them.

Virtually all slow-moving creatures use chemicals or poisons in combination with darts, spines and needles. Cone shells pack powerful, even lethal, poisons in their darts and regenerate them after every use. Both corals and sponges sport needle-like spicules, and both are often poisonous. Sea jellies, corals, anemones, and hydroids use a mass of minute harpoons to fire stinging nematocysts, although the potency is variable and suited to their particular needs.

If you doubt the virulence of stingers, observe the evidence around sea anemones. Little else grows around them for as far as the tentacles can reach. If the anemone cannot find a suitable substrate, it sometimes attaches itself to stony coral plates where it must battle for living space. The measure of its own nematocysts against those of the coral soon shows: a large outer rim forms around the anemone, almost as if the coral has pulled away in horror. Additionally, the coral forms a thick calcareous scab around these borders to protect itself from further damage.

Where a wide variety of coral competes for space, confirmation of the imperceptible battles between neighbours can often be seen. Corals use their stinging ability to repel any encroachment by adjoining neighbours and damaging grazers. Sometimes the borders that they build are small and friendly, but when corals are potent and determined to remain equally dominant, frontiers between them are thick and strong.

Above: *Cone shells* (Coniidae) *use poisons for defence. Some can even be lethal to humans.*

Some nudibranchs (or sea slugs) ingest and appropriate the stinging cells of reef organisms, somehow preventing the discharge of nematocysts, and store the stolen weapons for defence. Their other relatives use chemicals and toxins to impart distasteful properties. They all loudly and clearly advertize the fact with brilliant warning-colour displays; the colours may differ, but the 'hands off' message remains the same. Even sea cucumbers exude poison into surrounding waters or have poisonous mucus that can also severely affect divers' eyes. Soapfish, some gobies, toadfish and clingfish all excrete skin poisons, as do the Moses and peacock soles. Boxfish, defensively packed in little bony crates, can sail over reefs undisturbed. They emit poisons too, and in addition, join pufferfish in storing lethal poisons in their livers, making them deadly dangerous to eat.

In fact, the use of venom combined with spines is one of the most widespread methods of defence. The notorious stonefish is perhaps the most feared and best known example. For animals that move very seldom, they are deceptively fast when necessary and a brush with this master of disguise causes agonies that would fit into Dante's *Inferno*. All relations of the scorpionfish – including leaf fish and waspfish – carry potent venom glands at the base of their dorsal, ventral and anal spines that, when pressured, function as effectively as the modern syringe. The magnificent fins of the lionfish are also loaded with poison. Their splendour not only conceals the nature of this hunter but also prevents larger predator species from attempting to interfere with it.

Below: *Merlet's scorpionfish* (Rhinopias aphanes) *employs pattern and transparency to baffle all but the most practised eye.*

Above: *Fire urchins utilize brilliant colours to advertize that they are highly venomous.*

Below: *The masked pufferfish* (Arothron nigropunctatus).

Other species are less known for their use of the venom-and-spine combination and can catch divers unawares. The prominently striped catfish and some rabbitfish, as well as the very toxic crown-of-thorns star, are also equipped with dangerous poisons. The fangs of tiny sabretooth blennies are poisonous, hence these creatures are often mimicked by harmless fish. The barbed sting of the stingray is found almost at the end of its whiplike tail and can inflict a deep and painful wound. Bristle worms use irritating tufts of hollow, glass-fibrelike spines that burn like fire on contact.

Sea urchins are protected by venomous spiky needles as well as by pedicellariae – an arsenal of jaws, hooks, tridents, and venom glands. In everyday life, these are simply used as cleaning instruments. The flower urchin (*Toxopneustes pileolus*) is one of the most venomous in the sea; simply handling them can cause major discomfort and pain and, it is said, even death. There are many more species with venomous spikes or pedicellariae, and the utmost care must be taken with all, especially if they display brilliant colours. Toxicity can and does drastically increase during certain seasons. What may have been quite mild suddenly becomes extremely virulent. Divers must therefore always be alert and cautious around organisms which avail themselves of toxins as defence. Most of the poisons are protein-based neurotoxins and many cause excruciating pain and, in some cases, death.

In comparison to these formidable weapons, we forget other effective armaments. Surgeonfish carry razor-sharp caudal blades at the base of their tails. Cheek spines are used by squirrel- and angelfish. Triggerfish use their trigger to remain locked in crevices or even in the throat of predators. Barracuda and sharks have sharp teeth and streamlined efficiency that makes great speed possible. Porcupine pufferfish use spines and can ingest water or air to inflate themselves into threatening or inedible burred spheres. They can also bite severely. However, when they are inflated, they move clumsily and some time is needed to expel water and return to normal, swimming size.

The creature most maligned by inexperienced divers is the eel. Normally it is shy and almost puppyish, interacting easily with divers. When food is introduced, the equation changes. Eels have notoriously bad eyesight but a keen sense of smell – fishy fingers smell and look like food and so get bitten. Hands poked into dens, accidentally or not, are a direct threat and will be treated as such!

The tiny mantis shrimp not only has sharp tail spines, but also powerful extension claws which are capable of splitting a tough shell as well as an interfering diver's thumb. The electric ray manufactures electricity and can impart a 100–200V shock. The strength of the jolts is carefully calculated: some serve as a warning, others stun prey into a stupor or cause death. Snapping shrimps use explosive sound, angelfish loud deterrent drumming, to declare dominance in their territory.

Several creatures use distractions, casting off specialized appendages or emitting sticky substances or even parts of their gut in a more passive but still effective defence. Octopus, like cuttlefish and squid, use smokescreen fluids, but go a step further and instantly pale behind the fluids to escape, either by jetting away or by melting into the reef. They also possess fearsome parrotlike beaks and poisonous saliva. The blueringed octopus, though tiny, is lethal – the poison of its bite causes death.

In spite of all these formidable weapons, however, there is incredible restraint; under normal circumstances, the natural reef is not a dangerous place. Enter the diver, in marine terms a clumsy invader; yet there is not a single creature that will attack him unless it is extremely provoked. Most animals simply retreat out of reach. This even includes almost all species of shark, for they give clear posture warnings when their space is threatened or invaded. Under natural conditions they are shy and difficult to find; we have to search for them, rather than be approached by them. Almost all shark attacks are caused by erratic behaviour or actions that are perceived to emit prey signals. Of all human swimmers, the diver is the least endangered, mainly due to the lack of turbulence his movements generate. Of course, the introduction of speared fish or bait instantly initiates the feeding impulse and will certainly cancel any perception of him being an innocent part of the sea life.

If we simply recognize and understand Nature's weapons, we will be able to dive among marine creatures perfectly safely. The truly remarkable restraint displayed by reef life, except in life-threatening circumstances, deserves our respect and admiration. The same restraint should also set an example for our own behaviour.

Above: *The sabretooth blennie* (Plagiotremus tapeinosoma), *unlike the cleaner wrasse it imitates, has an underslung mouth.*

Below left: *The squirrelfish* (Sargocentron spiniferum) *displays his cheek spine.*

Below: *The bluespotted stingray* (Taeniura lymna).

Seeking the Sea's Creatures

Developing the Mindset

Previous pages: *Hermit crab* (Dardanus *sp.*).

Below: *Close proximity allows a better study of marine creatures. Be careful, however, that neither you nor any of your equipment touches the reef.*

The explorer who is interested in more elusive marine creatures must develop a special conscious mindset. Instead of racing over reefs to 'see it all', divers should concentrate on smaller sections of the reef to 'see much more'. The eye must be trained and developed to constantly separate the individual from the multitudes. Professional film makers scout reefs several times on submersible vehicles and make decisions long before taking their cumbersome equipment down. By all means copy the system of scouting, especially if you are able to dive the same location several times.

Of course, air limitations and safety rules force scuba divers to restrict their bottom time. Thus, superficial sweeping over reefs is usually the most that beginners, who still gulp lots of air, can manage. At this time it is important to get to know the general layout of tropical reefs in the broad sense. As both diving and observation experience increases, begin looking for different layers of life – sections where specific creatures habitually seem to occur. This is the important beginning of what

will become a preconceived dive plan on future dives. The whereabouts of specific animals can soon be quite accurately guessed at because, although the particular species may differ on reefs that are continents apart, the typical ways of every genus remain consistent.

Most seasoned divers know that continuously rewarding encounters are only obtained by careful scanning of small parts of the reef. A calm and easy-breathing diver lasts much longer and allow himself the time to see and experience much more. The additional advantage is that, once settled, he soon becomes a curiosity on the reef rather than a cause for alarm. Without obvious stalking, reef life quickly resumes its normal pace. In turn, these divers are then able to focus on minute movements that are often all that betray camouflaged animals. They develop that honed instinct for scenes that appear superficially normal but somehow just do not 'feel right'.

The external appearance of plants and animals reveals a great deal about how they have adapted to reef life. If we remember this we have the advantage of mental picture clues beforehand which keep us alert and able to register detail much faster. Fish are streamlined, clearly conveying mobility through their muscular bodies, their fins and the forward position of their eyes and mouths. Relying on great speed to capture prey and avoid predators, fast forward-propelling open-water species have indentations into which side fins fit for streamlining, and almost always tiny triangular turbulence spoilers along either side of their tails. Their muscular, bullet-shaped bodies and deeply indented tails differ greatly from the more flattened shapes and straight or rounded tails of slower, but more tightly manoeuvring reef fish. Flattened bodies indicate the ability to hide between corals and in narrow cracks and crevices, while the pectoral fins of hovering fish are very different from those of swimmers.

Above: *The clown anemonefish* (Amphiprion ocellaris).

Below: *Small sections of the reef can keep divers busy for an entire dive.*

Right: *Spotted sweetlips* (Plectorhinchus chaetodontoides).

Bottom: *Crocodilefish* (Cymbacephalus beauforti) *are typical ambush predators.*

DEVELOPING THE MAGIC EYE

Practise 'seeing' and pay close attention to details of external anatomy. Watch carefully for typical and repeated movement strategies or reactions. If you are a photographer, this will help you guess where your subject might move or what it might do next. Read about and memorize species behaviour and habitats. Very often an animal can be found simply because you are able to calculate where it should feed or live.

Play a game of marine 'picture this' by looking at photographs and absorbing details. Then cover them up and make a rough sketch of body shape and decorations. Do you fail to depict markings correctly? Do you notice which lines run solidly through the bodies and which stop short of the edges? How many bands did you count? Was there contrasting colour on the tail or dorsal fins? This information is vital for identification. Use the game to make yourself some pocket money by betting with buddies at your dive club. Few of your diving friends will have learnt to 'see' in detail. Young divers whom we have coached have financed many a coveted diving gadget in this way. By honing your skills, you will be able to describe and identify marine animals like a professional!

The shape of a creature's mouth is a very good indication of how and where it lives and feeds. Long snouts hint at plucking and picking on polyps, mucus and tiny organisms like crustaceans which live on sponges, corals and algae. Fleshy lips indicate sucking of succulent snacks. Powerful jaws do the crushing and grinding. Long sharp teeth and eyes set well behind the mouth feature strongly in fast fish-feeding predators. Bottom-feeders have mouths that face downwards and some may sport additional sensory organs for foraging in the sand. Ambush predators have large, upward-facing mouths with which to trap their prey. Slow-moving and sessile creatures, in contrast, are not streamlined at all. Many of them are radial and have effective defensive armaments. Their feeding options are much more limited than those of fish. Food may come from any direction and include a broader choice of plankton, algae and organic detritus.

Colour plays an important role underwater. In the presence of sunlight, brilliant and piercing colours are profuse. But as you descend deeper, increasing numbers of creatures are coloured in various shades of red. Red is also the first colour to filter out at depth, so take note of how creatures employ the colour. The stealthy hunt of a coral trout is assisted by jewelled speckles which help it blend with the reef, but its red body helps it to melt into the shadows almost unseen. Many nocturnal hunters, like soldier-, squirrel- and cardinalfish, all sport tones of vibrant red. At night, red appears velvety dark, a colour that seems denser than black. Teeth-gnashing colour combinations scream: 'Watch out!' Bright colours warn: 'I might look appetizing, but I've got a trick up my sleeve'. Subtle dappling and stippling meld better with reef floors. To me, yellow is perhaps the most frustratingly puzzling and mysterious marine colour. It is lush, saturated and brilliant, even at depth, where it should have long since been filtered out.

If we try to find an answer after having asked: 'Why this particular pattern?', we soon begin to understand camouflage strategies much better. For an alert diver, every dive imparts bits and pieces of new and useful information, which eventually add up to a vast knowledge of the reef, until suddenly we are able to imagine and intuit unknown strategies that may be used by the smallest, the rarest and most successfully concealed creatures.

SCIENTIFIC CLASSIFICATION

Once you become fascinated by marine creatures, you will inevitably want to identify, read about and discuss your discoveries. It is for this reason that an understanding of scientific classification is essential.

Descriptive common names are local creations. However easily they roll off the tongue, they are absent in the indexes of international field guides and encyclopaedias, which classify by scientific names. In other countries and languages, common names for the same creature may be completely different. It also does not suffice to describe 'a yellow fish with black stripes and black spots' if you wish to identify it. There is a bewildering array of yellow fish species with black stripes and spots, many of which are perhaps only subtly different. Accurate descriptions are crucial.

Thus, it was to facilitate identification internationally that, in 1758, the great naturalist and scientist, Linnaeus, designed a system by which all animals or organisms could be classified. The logical criterion was basic body design.

Explained simply, all living things, plant and animal, are grouped into 'phyla' which indicate the common evolutionary origin of the organisms. There are finer divisions into classes and orders but these are not necessary for our purposes. Each organism's scientific name consists of two words – a system known as binomial nomenclature. The first word is the organism's genus and always starts with a capital letter. The second word starts with a lower-case letter and indicates the specific species.

Let us use the lined butterflyfish as an example:
The phylum is chordata, *indicating all animals with a notochord.*
The genus is Chaetodon, *indicating all closely related butterflyfish.*
The species is lineolatus, *the Latin word for 'lined', indicating this particular species of butterflyfish.*
Thus the scientific name is Chaetodon lineolatus.

Learn to understand and use the system gradually and you will soon be better equipped to quickly identify those intriguing inhabitants of the reef.

Above: *Masked butterflyfish* (Chaetodon semilarvatus) *usually occur in pairs and mate for life.*

Above left: *Forster's hawkfish* (Paracirrhites forsteri).

Left: *A pair of delicately fringed ornate ghost pipefish* (Solenostomus paradoxus).

TOOLS FOR THE TASK

Herewith are some extremely useful tools for underwater exploration:

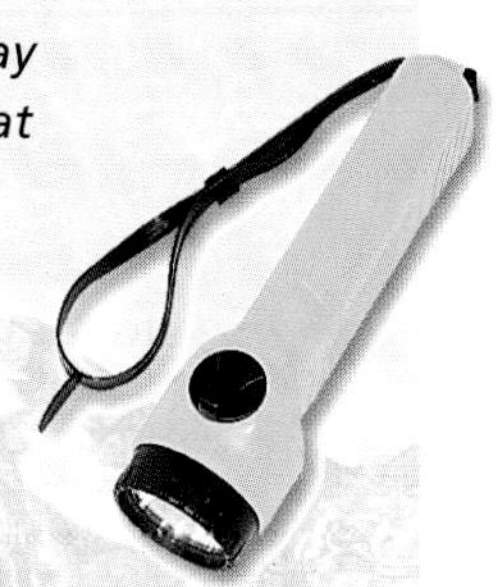

- **A small submersible light** – *indispensable and equally useful both day and night to reveal the true colours of marine creatures. Remember that colours are progressively filtered out as you descend below the surface. Initially red disappears, then orange, yellow, green, blue, indigo; finally, violet is lost and darkness begins. Back-lighting will enable you to see the tiny hairs of fire corals, the true colour of soft or gorgonian corals, and will accentuate magnificently translucent polyp pinnules. In addition, lionfish always investigate lights, while eels seem to be fascinated by them, and they will often pop much further out of their holes when they see a fixed light source.*

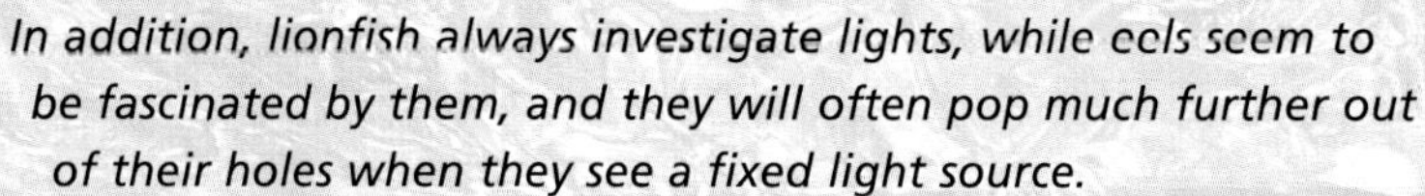

- **A magnifying glass** – *enables you to examine tiny features or fragile detail, especially on coral polyps and small creatures like sea slugs or flat worms.*

- **A white waterproof slate** – *handy for notes and quick drawings, as well as for messages to your buddy. It also effectively reflects light into shadows if you are a photographer.*

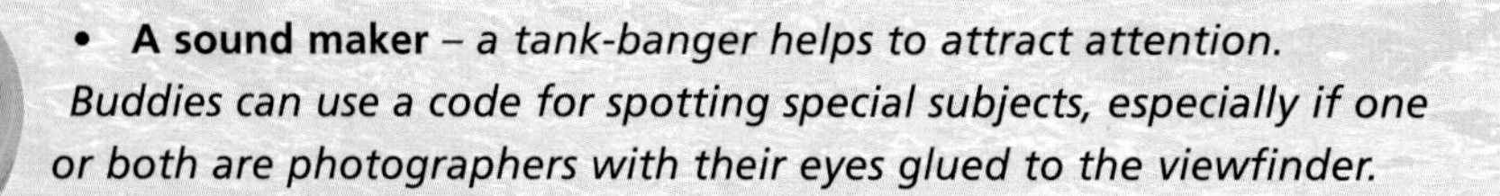

- **A sound maker** – *a tank-banger helps to attract attention. Buddies can use a code for spotting special subjects, especially if one or both are photographers with their eyes glued to the viewfinder.*

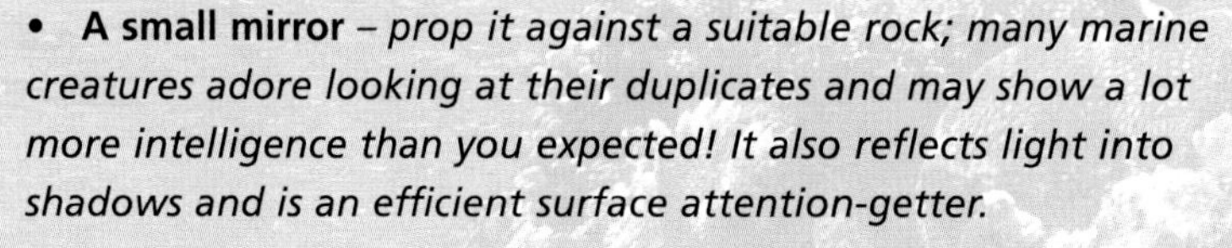

- **A small mirror** – *prop it against a suitable rock; many marine creatures adore looking at their duplicates and may show a lot more intelligence than you expected! It also reflects light into shadows and is an efficient surface attention-getter.*

- **Equipment clips** – *keep equipment, hoses, regulators and the like neatly in place and make yourself the most elegantly streamlined diver around.*

- **Cable ties** – *temporary fin straps can be fashioned, broken zipper loops replaced and other quick underwater repairs made to equipment like BC jackets, weight belts and camera lights or filters.*

LIVING TOGETHER

Each single ledge, each overhang, each tiny cave and all of the attached or slow-moving life forms are potential homes for some of the coral reef's most astounding animals. We would be fooled time and again if we did not remember that exotic creatures disappear amidst the jumble of visual images that make up the reef and clutter our minds and our ability to see.

To hedge the odds in our favour, we must learn which marine animals live together. Amazingly, it is not the antagonisms that we most see on the reef, but the alliances. Nothing is more central to coping with the challenges of life than cooperation – vulnerability is changed to strength and resources are efficiently utilized. These relationships of mutual tolerance or active cooperation are termed commensalism or symbiosis by biologists. In parasitic commensalism one member benefits but harms the other. In the more classic commensal relationships, different organisms live together for their mutual benefit. Sometimes they do not exchange benefits directly and only share proximity for protection or security, but essentially, the idea is to coexist, without either partner being harmed.

Above: *The commensal shrimp* Periclimenes soror *is frequently found on sea stars, in this case* Nardoa novacaledoniae.

Often, the chosen host species are those static and attached animals that form the living 'carpet' of the reef. They have many plantlike features that tend to belie their animate being. A general rule of thumb is that it is probably an animal if it does not resemble seaweed or grass. Venomous hosts or those that can sting are the most popular. So remember that sponges, anemones, sea stars, sea urchins, clams, sea cucumbers, sea fans, and the many various corals may all have one or more commensal partners. Most of the time, the commensal relationship with these hosts affords shelter or an occasional pilfering of food. The most interesting commensal arrangements involve more obvious mutual cooperation; the sea anemones with their clownfish are perhaps the best known example.

Commensal creatures often assume the colours, posture or behaviour of their host. As incongruous as these striking patterns may seem in isolation, in their natural setting on or inside their host they become exquisite camouflage. Remember, therefore, that it is the various characteristics of the host that hold the key and provide the clues for finding the sometimes tiny commensal partners. The shape and colour of the host body are the first clues, the shape and colour of its appendages the second, and the individual body patterns or markings, the third.

Once we look for the correct 'ingredients', we see these exquisite creatures much faster and sometimes even wonder why we missed them before. This cultivated instinct, with practice, soon becomes natural and turns us into successful 'spotters', able to locate rare and exotic species on reefs which may be thousands of kilometres apart, and on dives which others find boring.

TYPES OF COMMENSALISM

In all cases a commensal relationship involves at least two main parties. One of the partners will play the role of host. The other, which may be one or more organisms, fulfils the role of guest or lodger and is called the symbiont or commensal. A simple, protective arrangement in which the symbiont or commensal lives with a host – but need not – is called facultative. Barring the convenience of protection, nothing further is contributed or detracted by either party. A more complex arrangement in which the lodging organisms are individually or collectively survival-dependent on the host, is called obligative. In this case, the lodger has adapted so totally to the host that the situation has become permanent. There may or may not be additional swapping of services. However, while the hosts in this case may occur without their commensal inhabitants, the obligative commensal inhabitants will never be found without the hosts.

Left: *The body tufts of ghost pipefish* (Solenostomus paradoxus) *mimic their soft-coral host's polyps.*

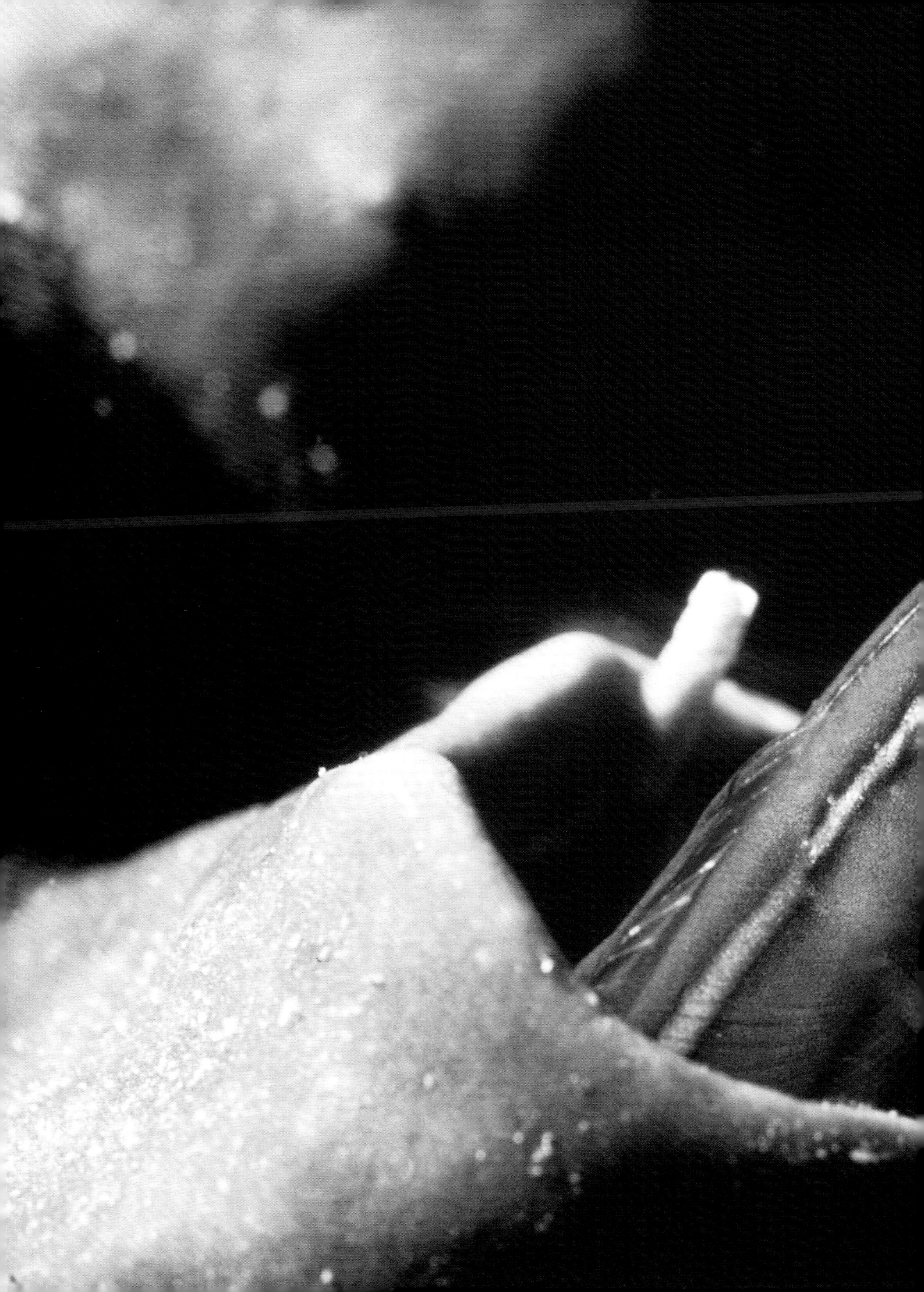

Habitats and Homes

Soft Coral Condominiums

Of course not all corals build stony skeletons. Many are soft and flexible with only a few sharp needles scattered within their bodies to help stiffen and stabilize them. The prickly-feeling, true soft corals, *Dendronephthya*, have beautifully saturated colours: purples, reds, yellows, oranges, pinks, whites, and blues. They, like all other corals, utilize the standard design of the sea anemone, the polyp, to build their beautiful colonies and capture carnivorous prey.

Each soft coral is a veritable super-condominium for creatures. Inspect them gently and be prepared to hang around quietly to pick up minute movements. Shrimps on soft corals may be as transparent as glass, sporting just a few spots to match markings present on the host. Others are spattered with intricate patterns imitating the surface of branches or the pattern and colour of their specific host's polyps. Only the claws of others take on the coral colours or sport highly contrasting colour at the tips of their pincers. Look for the very tiny Dendronephthya crab (*Holophrys oatessi*) which is almost indistinguishable from the coral, and the arrowhead crab (*Huenia heraldica*). But remember that arrowhead crabs and sea spiders, like most other crustaceans, are best seen when they emerge at night.

Decorator crabs that live on soft corals are smart. They match their bodies to the main trunk of coral, but then pinch off live coral polyps which they hold in their claws or attach to their bodies, successfully faking polyp sets or clumps. They frequently assume a posture that matches the structure and

Previous pages: *The sabre-tooth blennie* (Plagiotremus tapeinosoma).

Right: *The decorator crab* (Naxiodes taurus) *plucks and 'plants' live polyps on his carapace to create a camouflage costume.*

Below: *Look very closely for allied cowries* (Phenacovolva rosa), *as their mantles perfectly imitate coral polyps.*

build of parts of the coral and almost always align themselves with connecting branches or at those spots where their decorated carapaces will most logically and successfully mimic a clump of polyps. Nudibranchs that eat specific corals will match their tufts and appendages to the polyps, their bodies more often to the trunks. Tiny, big-eyed coral gobies (*Pleurosicya mosambica*) are almost transparent; the spattered markings on their transparent bodies match the gel-like translucency of the coral stem and its surface texture. Although they usually sit quietly on the coral trunk among the spicule patterns, their large dark eyes or a nervous wiggle may betray them.

Many shell species, but particularly the allied or spindle cowries, live in association with a variety of soft corals. Be aware beforehand of the different colours of mantle and shell, as this will prepare you for the convincing ploys shells can employ. The mantles of the *Prionovolva* shells are almost indistinguishable from the corals they graze on, as the colour and texture match clumps of polyps almost exactly. These molluscs tread the line between parasitism and commensalism, for they graze on the surface of the host but apparently do little harm. The flamingo tongue shells (*Cyphoma gibbosum*), with their ringed patterns, live on lacier, more netlike corals. Although more obvious, they are often overlooked by the uninitiated because the pattern melds with the coral's tiny windows.

The toenail cowry (*Calpurnus verrucosus*), which is not a true cowry, lives only on a specific soft coral (*Sarcophyton* sp.). Very different to *Dendronephthya*, this host is greyish-white and leathery to the touch and constructed like a single, wavy, floral cup on a stalk. Sometimes solitary, it more often occurs in colonies where the single 'blooms' mass tightly together. The translucent white shell of the toenail cowry is offset by a delicate mantle and foot that is spotted to resemble the pattern of the minute cuplike calices of the retracted polyps. The mantle completely covers the dainty white shell by day, providing excellent camouflage.

The golden wentletrap shell (*Epitonium billeeanum*) is more often seen at night when it hides between, and feeds on, golden *Tubastrea* and *Dendrophyllia* corals, on which it also lays its matching golden eggs. By day the translucent polyps retract, leaving only bunches of pretty pink pipes.

Above: *Sea pen blades become multistorey condominiums for tiny glassy shrimps.*

Left: *The patterns and translucence of the hingebeak shrimp* (Rhynchocinetes *sp.*) *help it to disappear among soft corals.*

LIFE IN LATTICES AND LACEWORK

Above: *The depressed crab* (Xenocarcinus *sp.*).

Sea fans, sea whips and gorgonian corals are horny corals. Although animals, they look like the shrubs and trees of the reef and may be white, cream, grey, yellow, pink, orange, deep red or even blue. All tend to stretch out from the edges or flanks of the reef, seeking the best vantage points for collecting food. Their position is dependent on and always at right angles to the flow of prevailing currents. This allows them to strain as much as possible from the passing plankton bouillabaisse.

Their branches are covered with large collections of the basic polyp which in turn are all connected to a single nervous system with minute filaments woven through microscopic internal pores and canals. If you gently touch a part of the branch, some of the polyps will retract in both directions from your touch. Gently finger them at the fork of a branch and the retraction occurs over a much larger area, on both sides and beneath. This collective reaction confirms that they are not plants but animals which should be treated with the utmost respect and gentleness.

The horny corals hide plankton- and tiny crustacean-feeding creatures. Crustaceans, like crabs and shrimps, are abundant here. Gobies and spindle-shaped allied cowries also settle on them and have adapted their body shapes to match their hosts. The gorgonians are netlike, with strong main branch stems on which bodies can be hidden and eggs attached. But the open 'window' patterns between their lacy blades may also be cleverly used as illustrated by the dainty longnose hawkfish (*Oxycirrhitus typus*). This fish has developed red lattices on its silver body to imitate both the solid connecting structures and the open gaps of the gorgonian fan. The individual tartan pattern, combined with the fish's ability to sit very still, affords a convincing disguise. Where it is denied this space, perhaps due to lack of gorgonians or to overpopulation, it lives and is equally well hidden in bushy black corals. Here it adopts a more subdued orange-red. Patience and a gentle approach are required to spot it, but do not to bump the coral. Once alarmed, the fish dart around, making photography almost impossible. Photographers should prejudge the distance, retreat and 'lock' focus on a similar distance elsewhere, only then returning, ready to 'click' at the perfect moment.

The beautiful depressed crabs (*Enicarcinus depressus*) live normally in pairs and only on gorgonians. Their red colouring varies in intensity to match the specific coral on which they live. They have an opaque white stripe down the length of the back. Also look closely for Anga's ovulid (*Phenacovolva angasi*), a longish spindle-shaped shell with an exquisite red mantle spotted in white, on red gorgonians which extend white polyps. Found only on these gorgonians, Anga's ovulid is truly difficult to see. Another spindle-shaped creature, the gorgonian shrimp (*Tozeuma armatum*), is equally obscure in spite of its 6cm (2in) length and blue-and-white mottled stripes. Exquisite macro-photography subjects, colonies of the rare striped sea anemone live only on gorgonians and are more easily seen.

Black coral shrimps (*Periclimenes psamathe*) may be betrayed by their white eyes. The juveniles are completely transparent, while the 3cm-long (1in) adults have a longitudinal red stripe on the back, which may look green or black in natural light. The dense growth of the bushy branches offers a maze of hides so they are not easy to find, especially when polyps are extended.

The 'black' of black coral refers only to the inner skeleton; the live colonies are never black when seen underwater and may be coloured yellow, white, brown, orange, or green. The black coral crabs (*Xenocarcinus conicus* and *Quadrella granulosa*) have complementary yellowish or ochre bodies, which blend with similarly coloured polyps. One of the clinging snake stars (*Astrobrachion adherens*) is found only on black coral trees and it may display one of two different colour patterns: a brilliant orange-red or a more subdued brown.

Following pages: *A longnose hawkfish* (Oxycirrhites typus) *seeks refuge, matching its gorgonian home.*

Below: *In its lacy coral home the translucent goby* (Gobiidae *sp.*) *is securely hidden from prying eyes.*

Above: *Wire coral should always be inspected for a variety of inhabitants.*

Sea whips and wire- or corkscrew corals also extend into the currents, wire corals sometimes by as much as 3m (10ft). Again, they occur in a full range of colours. The whips, harps and wire corals clearly indicate what body shapes may be expected on them: slender gobies and short-legged shrimps as well as very small crabs and spider crabs. Sometimes the gobies' quick dart to catch plankton delicacies might be an early indication of their presence but, once they are aware of you, they will remain very still or duck behind the tendrils. If you cannot pinpoint them, wash the beam of your light across the coral a few times to induce movement. They usually live in pairs and at times attach their eggs to a prepared patch of substrate along the coral's length. Here the eggs are cleaned, aerated and guarded until they hatch. These gobies are tiny and translucent and may be deceptively blotched. Their relatively large, dark eyes are usually the first obvious feature visible in the beam of an underwater light.

On black coral whips you might find the Zanzibar shrimp (*Dasycaris zanzibarica*), which lives off plankton and mucus from the whip. As the male is a dark ochre, it will only be found on a similarly coloured whip; the female is much paler in colour. Both have irregular, white, side-to-side stripes and sport a single large nipper.

Old wire corals are often colonized by feather stars, encrusting sponges, various ascidians, delicate glasslike tunicates, and small, feathery, fernlike hydroids in a host of colours and shapes. These growths turn them into beautiful stems that look like fruit-laden branches. Exquisite miniature, posylike arrangements, they make super macro-photography subjects. Wire corals are also populated by shrimps, nudibranchs and flat worms.

Although hydroids on their own are not as spectacular as the soft corals and gorgonian fans, they add delicate beauty to the reef. Hydroids are feathery animals that look like translucent plants and they flourish in all tropical waters. The majority grow in low-profile clusters of fine structures, but

EVER-CHANGING CORALS

Remember that the overall appearance of all soft corals can change drastically, depending on whether their polyps are retracted or extended for feeding. Occasionally even experienced divers are fooled, for the change can be so remarkable that a reef seen in slack current might be unrecognizable later when the current runs. Likewise, remember that corals grow largest and are most abundantly populated where currents are periodically strong.

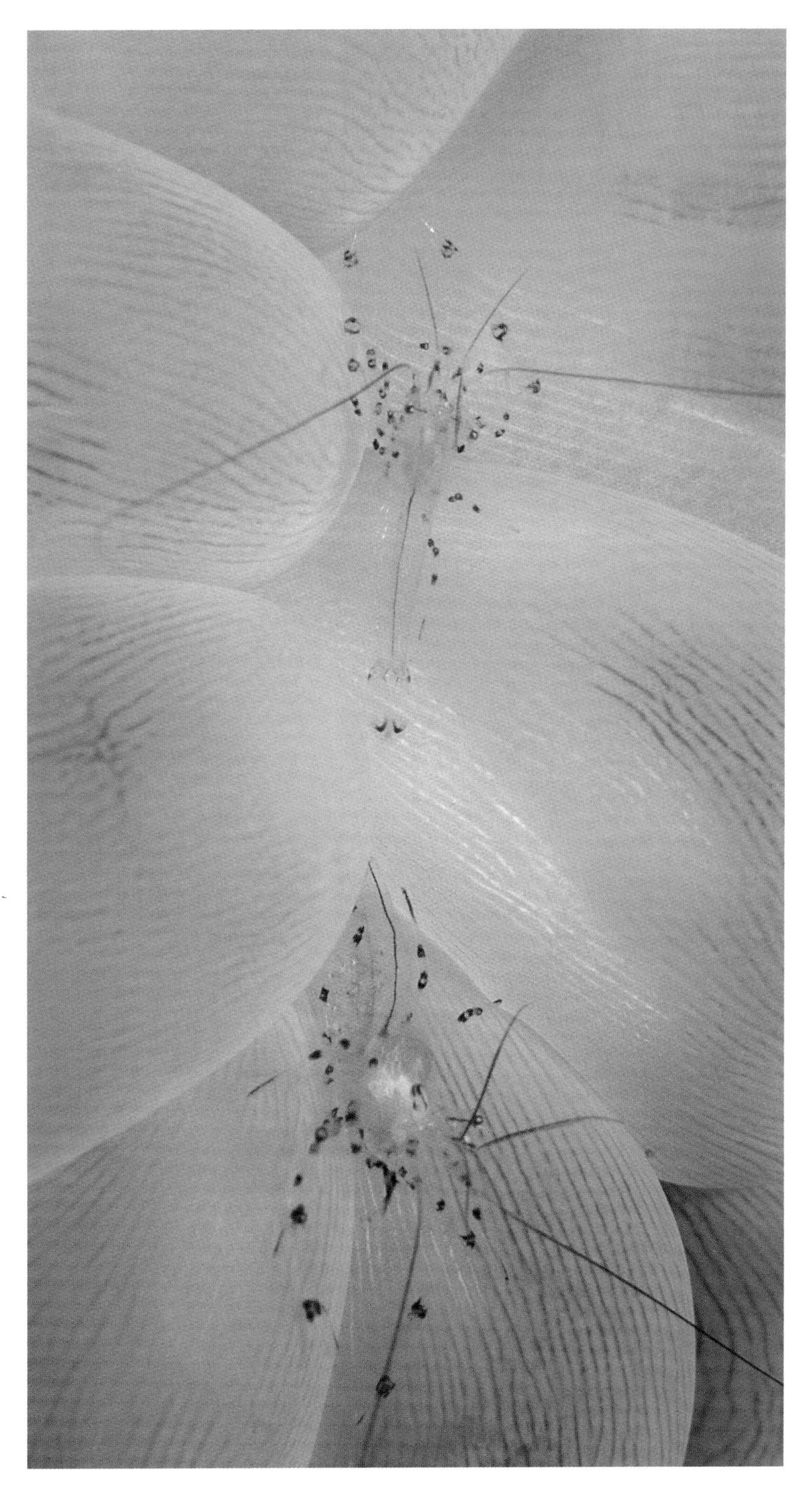

at night. Coral hermit crabs live permanently encased in hard corals. Bubble corals are profusely populated by numerous species of shrimp, many of them glassily transparent. Lobsters and slipper lobsters hide in and under dark overhangs.

Boxfish, pufferfish, pipefish, a variety of juveniles, and several nocturnal species seek shelter among coral bases. Branching corals offer shelter to fish species such as damsels and cardinalfish, the rarer filefish and shrimpfish, as well as the juveniles of angelfish, sweetlips and barramundi cod.

Solitary mushroom corals are hard, stony and multibladed, shaped like the underside of the fungi they were named after. They occur on reef floors often near sandier patches. Some push out so many polyps at night that the whole body is obscured. One particular species in the south Pacific, the sunloving coral, resembles anemones. Many shrimps that normally inhabit anemones are also found in these corals.

Do remember that all stony corals are living animals. Refrain from touching them, for they are easily damaged and will in desperation secrete copious amounts of slime to get rid of what is damaging them.

RESIDENTS OF THE REEF FLOOR

Deeper down, bereft of the tapestry colours of the reef, the floor seems a vast and bleak desert devoid of life. In reality, it is filled with treasures and, while they are much more subtle in colour and pattern, they are no less unique than their vibrant neighbours.

Although all fish seem to be aware of the delectable roe held within urchins, few predators actually bother them. Their globular spine-covered bodies are reasonably impenetrable. They squat motionless in hollows during the day, but at night clamber out and perambulate nimbly on their tiny pedicellariae to forage. The urchin with its tiny spiny forest makes an excellent shelter for young fish, who can dart between the spikes for protection. When the urchins tuck into protected, dark hollows, exquisite flagtailed pipefish soon share the space.

The hingebeak shrimp (*Rhynchocinetes hiatti*) is a spectacular night species that can grow up to 8cm (3in) long. It lives in loose association with sea urchins, and hides behind them during the day. This shrimp is not uncommon but it is quite elusive. The needle shrimp (*Stigopontonia commensalis*) is a prime example of how body shapes change to benefit from commensalism. Specialized to live only on the black long-spined sea urchin, *Diadema*, this shrimp has grown long and spindlelike, with lengthways white lines that blend with the urchin spines. The shrimp has developed shortened legs that are adapted to grasp the thin spine needles and, in addition, holds its claws closely aligned with the spines and its body.

Another creature that is encountered exclusively on the common urchin is the urchin crab, *Zebrida adamsii*. With its distinct zebra pattern, it is well camouflaged among the dorsal spines, otherwise it seeks shelter beneath the body of its host. It holds onto the spines with modified hooks at the end of its legs. The crab is almost always found alone, mostly in shallow-water urchin colonies.

Below: *Reef floors seem barren, yet are home to a multitude of exotic creatures.*

The decorator crabs are perhaps the most accomplished artists on the reef floors, often able to grow delicate fernlike hydroids on their hard body shells and meld with their surroundings. Some are so intensely obsessed with the merging strategy that they will actively 'plant' and encourage the growth of different marine organisms until not a spare millimetre of the carapace is left uncovered.

The shy but exquisitely coloured reef lobsters (*Palinurella wieneckii*) are vibrant red and magenta. Slightly larger, around 7 to 10cm (3 to 4in) long, they maintain dens in crevices and under small overhangs. Very secretive, they are super-cautious when alarmed and thus not easily found by noisy divers. Rock lobsters (*Panulirus versicolor*) are much larger and lurk inside their dens more often than not during the day. When it is properly dark, they emerge for hunting food and scavenging and can then be seen more readily.

The picturesque harlequin shrimp lives on or close to reef floors too, where it can easily find the sea stars on which it feeds exclusively. It is tiny and secretive, so try to find it by inspecting the more conspicuous stars. The fascinating mantis shrimp (*Odontodactylus scyllarus*), with its cheeky personality, thrives in the reef-floor domain; very often it chooses the more rubbly areas for burrowing its well-camouflaged home. These shrimps are extremely territorial and can often be seen at the burrow entrance, totally caught up in their meticulous toilette. This is one occasion when a tiny titbit placed near the burrow or a small obstruction over the entrance might lure them out momentarily, but they will return to the protection of their house in a flash, so photographers should prepare camera settings and lock focus in advance. They are voracious predators of other crustaceans, small fish, molluscs and worms and use their powerful dactyls (specially adapted front limbs) to smash open the outer shells of their victims with tremendous force. The bullet-speed with which they do so is recognized as one of the fastest animal movements known to humankind. Do not be tempted to place challenging fingers or clumsy thumbs anywhere near them – you will receive blue, painfully throbbing bruises! Sometimes mantis shrimps cockily stroll across the reef and if not alarmed, might engage in a tenacious staring match with you, assessing whether you are a possible threat. Remain still, for if you pass the test they will merrily carry on their way.

All tube anemones should be closely inspected, both at night when they trail their long tentacles into the currents and during the day when they are retracted into their tubes. They are a favourite hideout for shrimps, especially if the tube extends well above the floor surface. The tentacles may

KILLER CONES

A word of warning for the uninitiated: cone shells are carnivorous killers and some even aggressively hunt fish. For this they are equipped with extremely venomous darts which are constantly regenerated. An injection from any of these molluscs is dangerous and possibly deadly. The geographic cone (Conus geographus) *is the most poisonous shell of all and its neurotoxin is much more potent than cobra venom. Anybody interested in cone shells should make a detailed study prior to randomly touching or picking up any specimens.*

Above: *Study the mantle features of cowries to match them to their reef surrounds.*

Above centre: *Bubble shells* (Hydatnidae *sp.*) *are usually found on the reef floor.*

Above right: *Money cowries* (Cypraea moneta) *served as currency long before coins were invented.*

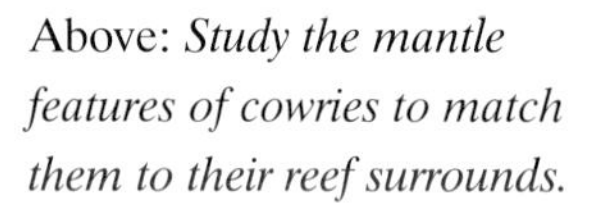

retract suddenly if the animal is disturbed. Also always keep an eye out for sea hares, shell-less molluscs that feed on these stinging tentacles. Your chances of seeing them are relatively small, but you may just be lucky enough to spot one!

The rubble-strewn or sandy patches of the reef are mollusc paradise. Many mollusc species are strictly nocturnal and can never be seen during daylight as, in almost all cases, they burrow into sand or under dead coral slabs, branches or boulders. The keen-eyed spider shells (*Lambis* sp.) and ploughing tun shells (*Tonna perdix*) are found on these patches, as are the giant helmet shells (*Cassis cornutus*). In the rubble one can also keep an eye open for brown or reddish tentacles that resemble those of anemones. If the rubble is carefully removed around the tentacles, a file clam or flame scallop will be found. These amazing bivalves scramble for cover in short, frantic, hopping bounds. The tentacles are a deep, rich red when lit, perfect for a stunning macro photograph. Videographers will enjoy the comical short hops with which the clam travels. Make sure that it is adequately hidden again before you continue, for it is coveted by carnivorous fish.

Often, the outer surface of spider and helmet shells is rather unattractively covered in camouflaging reef-matching algae, totally different from the glossy inside. It is astounding how easily these shells are overlooked, if one is unaware of the mimicry they employ. The

shiny cowries avoid detection by wrapping themselves in variously patterned, furry-looking mantles. These mantles possess shell-producing glands which continuously build and gloss the shell from both sides. To find cowries, one must know the mantle colours of the various species so that similarly patterned surroundings can be searched. For instance, the related white eggshell cowry, *Ovula ovum,* possesses a black mantle with warty orange stipples. It is almost always found where an identically patterned encrusting sponge grows. When spawning, cowry shells sit on their eggs, unlike other molluscs, and this is certainly reason enough not to move them.

The giant clams (*Tridacnidae*) filter-feed but also derive food from another source: they keep their valves agape because millions of food-producing zooxanthellae live in the beautiful exposed mantle. These plant cells are also responsible for the vivid and iridescent colours of the mantle frills. The mantle contains semi-transparent windows through which sunlight is concentrated towards the zooxanthellae. So good are the plant cells at food production that the clam apparently requires very little extra from filter-feeding. These giant clams are the biggest and best known

Below: *Very large clams* (Tridacnidae *sp.*) *can no longer close efficiently – negating their fame as murderous diver traps!*

Above: *Artificial light reveals the beauty of peppermint-striped sea cucumbers* (Thelenota rubrolineata).

bivalves on the reef although we surprisingly seldom seem to think of them as shells. They can live for up to 200 years and become so enormous that they are no longer able to close.

Clams are extremely photogenic and have a very special light-reflecting property. If the mantle is looked at from opposite directions and depending on the angle of your view, the colours change completely. Iridescent blue may appear a vivid emerald green from the opposite side. There are also light- and pressure-sensitive spots that take the role of eyes, so that clams retract when disturbed by divers. The gills of the animal are located inside it and their incredibly detailed construction can be seen through the open vent. The strange 2cm-long (0.8in) pea crab (*Xanthasia murigera*) sometimes lives on clams, although it is also found living with partner gobies and their shrimps. Its name is derived from the concave shell which sports a little bump, causing it to look like a miniature pea pod.

When these giant bivalves spawn, it is an event of volcanic proportions, truly a spectacular sight to behold! Clams are hermaphrodites; in synchronization with others, they first heave and squeeze to release sperm, then about half an hour later, they contract to release their eggs. The many legends that describe murderous man-trapping clams are absolutely unfounded. Those big enough to fit the bill are so heavy that they wince, rather than slam, closed. Besides, the larger shells only touch at the crests of the wavy points, leaving ample gaps in between. In any case, divers really have no business inserting parts of their bodies into these harmless molluscs.

Thorny oysters (*Spondilidae*) are also bivalves; they attach themselves to many reef walls. The frilly inner mantle grins a rather grisly-looking smile until too closely approached. Then the clam suddenly closes, for it has rows of tiny but acutely light-sensitive eyes. However, they are often encrusted with growths like ascidians, hydroids and sponges on which shrimps live and nudibranchs feed.

Sea cucumbers (*Synaptidae*, *Holothuridae* and *Stichopodidae*) are those normally rather unattractive cucumber-shaped bottom-dwellers that obtain nourishment by ingesting copious amounts of the rich, organic filament that coats sandy bottoms, or reef-bottom sediment and detritus. As they slowly crawl along, expelling their characteristic trail of sand-sausages, they scrub tons of sand clean of products that would otherwise foul the water.

Some of these sea cucumbers are beautiful and intricately patterned. In the New Guinea waters a stunning red-and-white peppermint-striped sea cucumber, *Thelenota rubrolineata*, lives on the reefs, clinging to corals and rocks. In the Indo-Pacific the pineapple sea cucumber, *Thelenota ananas*, bears beautiful star- and flower-shaped protrusions over its entire upper body, in between which it harbours shrimps and tiny serpent stars. The leopard sea cucumber is named *Bohadschia argus* after the mythical multi-eyed god, Argus, for it has many brown dots surrounded by concentric white circles. It is a suction-feeder and its mucus is irritating to a diver's eyes. When disturbed, some sea cucumbers eject long white sticky filaments to divert enemy attack. Many prodding divers have had to extricate themselves from this unsavoury mess. Others eject part or even all of their gut as sacrificial bait. While these body parts are eventually regenerated, the process burns energy needed for other purposes so divers should refrain from poking these creatures. There are also several sea cucumbers with plankton-feeding tentacles.

Many of these strange animals are popular hosts for a variety of specialized shrimps, often members of the *Periclimenes* species. Some synaptid species can reach lengths of up to 5m (16.5ft). On one of these, *Synapta maculata*, a rather repulsive-looking and feeling species, the spectacular imperial shrimp (*Periclimenes imperator*) lives commensally. These brilliant red shrimps are normally found in pairs and certainly make no attempt at camouflage. However, they surprisingly often go unnoticed, perhaps because we automatically avert our eyes from the ugly host creature.

Several sea cucumbers, like the pineapple (*Thelenota ananas*), leopard (*Bohadschia argus*) and spiny black (*Sticopus chloronotus*), may harbour the rare translucent pearlfish (*Fierasfer*) in their gut. The symbiont's body is slender and smooth so that it may easily slip into the anus of its sea cucumber host, sharp tail first. The fish only emerges at night so look out for the hosts and do not storm the sea cucumbers before having watched carefully from afar.

Slow-moving and rather clumsy, sea cucumbers must make use of synchronized spawning into the water. During springtime they can often be seen, posturing bolt upright, swaying from side to side for several hours, emitting spawn from a gland in the head. So far, research has led divers and scientists to believe that this posture may be used only for sexual emissions.

Harmless upside-down sea jellies are often found residing on shallower reef floors. They do not sting and, unlike other sea jellies, the tentacles are kept upwards because the zooxanthellae that live in them require sunlight in order to manufacture food for their host.

Below: *Feathermouth sea cucumbers* (Synapta maculata) *look repulsive, but sometimes host the exotic imperial shrimp* (Periclimenes imperator).

STING-PROTECTED HABITATS

Below: *The white blotches of the eggshell anemone shrimp* (Periclimenes brevicarpalis) *led to its common name.*

Corals, together with anemones and sea jellies, are animals that sting; they are also exclusively carnivorous. For feeding and defence they use stinging cnidoblasts, cells which contain a capsule called a nematocyst, an ingenious little barbed harpoon. This spring-loaded weapon is activated either by touch or a chemical stimulus to discharge, penetrate and envenom the target. As every nematocyst can discharge only once, many hundreds must do so to overwhelm the prey.

Coveted for the protection this ability can afford, all anemones and anemone-related species like corallimorpharians are eagerly occupied by shrimps. These shrimps are real masters of inconspicuousness. Some of the most beautiful are almost totally transparent; these are the glassy shrimps which always live in pairs. The male is the noticeably smaller specimen. Watch for subtle giveaways: a white colour spot on the back, appearing much like a tentacle tip, or delicate colour hues on the claws. The greenish Amboin shrimps, *Thor amboinensis*, and also *Periclimenes brevicarpalis*, sport many sharply defined white blotches which look somewhat like crushed eggshells; this has led to the common name of eggshell shrimps.

The Amboin shrimp is the least host specific and, apart from other coral hosts, lives on no less than seven of the 10 species of clownfish-harbouring sea anemone. It is easily recognized by its rather funny posture, for it holds its abdomen and tail in a vertical position, while rapidly flicking it up and down. All the shrimps normally associated with anemones are usually also found on various stinging coral species.

Corallimorpharians (*Amplexidiscus fenestafer* and *Discosoma* sp.) by day look like flat, short-tentacled anemones. At night the large ones close into pot-bellied vases, enticing creatures into what is a false and lethal protection. Once inside, the victims are unable to escape and are consumed, a tactic also used by carnivorous flowers. But tiny, specialized shrimps are able to live on the smooth outside skin by pinching onto it; the nippers cause minute folds, a dead giveaway of their presence to a sharp-eyed observer! Corallimorpharians sting virulently, particularly the smaller species which often carpet vast stretches on reefs, looking dowdy and harmless. When agitated, they emit copious quantities of long white filaments that contain thousands of really lethal stinging harpoons. In humans their sting can, in some cases, lead to severe muscle atrophy or serious contact dermatitis resulting in painful blistering and peeling of skin.

Hermit crabs (*Diogenidae*) have long abandoned their usual carapace for the superior armour of empty shells. This choice has freed them from the regular moulting process, as they can easily slip into larger homes as they grow. The crabs relocate at night, carefully measuring the proposed home and trying out the fit before abandoning the old. Divers and carnivorous marine hunters tend to ignore these empty and dead-looking shells, proving how well the concealment strategy works. On reefs close to civilization's litter, hermit crabs may even use small bottles or other suitable containers for homes.

Above: *To stay aboard, corallimorpharian shrimps* (Pliopontonia furtiva) *must pinch onto the outer surface of their host.*

Below: *The anemones used by the anemone hermit crabs* (Dardanus pedunculatus) *occur nowhere else on the reef.*

A specialized relative of the ordinary hermit crab, the anemone hermit crab (genus *Dardanus*), has become party to 'the Sting' in a rather unbelievable way. These fascinating creatures add specific species of small live anemones (*Calliactis* sp. and *Adamsi* sp.) onto their shell homes, the only place where these small stingers grow, extending and retracting their tentacles in the usual way. Thus, the hermit crab is afforded extra protection, while the anemone benefits from the crab's mobility and rather messy eating habits. I once observed such a crab handing a morsel up to his anemone! Some of these crabs seem to have an ego problem. A sense of one-upmanship compels them to overload their shells with several anemones until they can hardly move effectively. So they stagger across the reef with a rather inelegant waddling gait.

Not to be outdone, the 3cm-long (1in) boxer crab (*Lybia tessellata*) is brightly patterned and not inconspicuous at all. But it needs no camouflage, for it also makes use of small live anemones. Grabbing two in its claws, the crab waves its anemone boxing gloves frantically at the slightest sign of a predatory threat.

INHABITING TUNNELS AND CHANNELS

Sponges (Spongiidae) are perhaps the most subtle sedentary animals on the reef. From an evolutionary point of view, certainly nothing new has developed from them for about 200 million years, and today they still filter water for the food particles it contains.

Sponges have no muscles and no nervous system, but they do possess pores that lead to a maze of interior channels. These channels are lined with tiny hairs that beat to create currents which draw in ever more water. What is edible is absorbed, the rest is expelled in outgoing water. Sponges can grow to considerable size and mutate into an amazing variety of shapes and colours, ranging from shapeless blobs to fingers, tubes, barrels, fans, trumpets and vases or even thin velvety encrusting sheets. Sponges of the same species can have totally different appearances, so much so that even scientists can sometimes only identify them conclusively under a microscope. Many also contain poison or secrete stinging mucus. Some contain spicules and most have threadlike spongin which serves to retain their shapes – this is what we use in our bathrooms once the living cells have been killed and removed. Sponges are extremely susceptible to silting and have no way to rid themselves of the clogging blockage. Except for photographers, most divers mistakenly pay little attention to them.

Below: *Sponges occur in thousands of forms and have remained unchanged over the millennia.*

PARACHUTES OF THE DEEP

Translucent and bell-shaped, the ocean's true drifters throb through the sea with heartbeat pulses. Erroneously called 'jellyfish', these creatures are far more intricate than can be guessed at from those anonymous blobs that sometimes wash ashore.

Looking like small planets of protoplasm, sea jellies have little control over their dispersal, be it in huge colonies or as solitary travellers. Most bear an array of clear bell-shaped skirts adorned with trailing, food-catching tentacles. A few of these provide residences for pelagic juvenile fish or waterbed homes for juvenile trevally.

Other jellies and salps assume bodies that are reminiscent of glass fruits or Christmas angels and brighten their nights with bioluminescent rows of flickering lights. These, along with the rare 'Belts of Venus', are open-water species and are almost always only encountered by divers during decompression or safety stops.

Beware though, for even the tiny toy parachutes that seem so harmless can pack a punch. The Cubomedusae*, a little parachute with just four long tentacles, is definitely one to avoid, while the jelly* Chironex fleckeri *has a sting that may sometimes be fatal. And yet, these parasols of solidified water, in spite of their potential as stingers and irritants, are easily avoided by divers.*

They are exquisite photographic subjects and their mesmerizing movements can be videoed with great success. When they are cleverly side-lit, the resulting images rival those of the most exotic reef creatures.

Above: *A mastigias jellyfish* (Mastigias *sp.*).

Below: *Extremely difficult to spot, anglerfish* (Antennarius nummifer) *usually occur on or near sponges, which they imitate closely.*

Few reef inhabitants eat them, and those angelfish or filefish that do, prefer to take a bite only after sponges have been damaged and the inner flesh exposed. For this reason, sponges create an important niche for other creatures. The velvety outer skins and hidden passages also constitute ideal homes. Many sponges harbour the exquisite commensal zooxanthius corals. Small species of mollusc occur on them in abundance and crustaceans are especially prolific. Other creatures like brittle stars and ribbon worms live and feed in or on them and several nudibranchs and flat worms feed exclusively on particular poisonous sponges.

The most rewarding, but most difficult creature to find around sponges, is the frog- or anglerfish (*Antennarius* sp.). Slow, clumsy and covered with protuberances, these fish emulate amorphous blobs of sponge to perfection, matching lumpy shapes as well as colour. This fish is an ambush predator and sports an amazing little angling rod on its forehead in the shape of a tiny filament ending in a succulent miniature flagged lure. The lure is rotated and bobbed enticingly in front of its camouflaged mouth as soon as small fish swim nearby. Sitting incredibly still, the fish is only exposed by the tiny movement of the lure or its very small, peculiarly gemlike glittering eyes. Small chalky-white or yellow spots on the surface of the bogus 'sponge' are evident on almost all frogfish and should prompt you to take a closer look. The pectoral fins are used to 'walk' and have a strange handlike shape that reminds one of webbed reptilian feet. This fin shape may be discernible as an unlikely feature on an otherwise smooth-looking 'sponge'. The pounce and gulp of a frogfish is so fast that it appears as a blur, even on stopped-down film.

Communities of the Deserts and Meadows

Opposite: *Eeltailed catfish* (Plotosus lineatus) *graze in schools across the astonishingly nutrient-rich reef floor.*

Below: *The skin of a speckled sandperch* (Parapercis hexopthalma) *imitates the patterns of sand grains.*

The magnificent stature of tropical reefs so entices a visitor that the white and apparently bleak sand deserts are often ignored. And yet, they are an inextricable part of the tropical reef and its community. They are the product of the crumbling bones and corals, the dissolving skeletons, the gut-filtered excretions and the calcification of algae. If they seem barren, the strategy of these rippled plains succeeds, for in reality they hold and hide a myriad subtly coloured and mysterious creatures that specialize in camouflage of more delicate dimensions. In the absence of visible shelter, burying, burrowing and swimming in sand become the norm.

Here we see the spurts of sand clouds from miniature volcanoes created by the secretive ghost-nipper shrimp. We startle at the clear snapping explosions of unseen pistol shrimps. The chameleon prawn reigns in meadows, concealing itself by constantly changing colour as it moves. The various sand-dwellers are slender in order to move comfortably between grainy particles; when they dive into the deserts, they do so with astonishing speed. The sandy plains are enveloping havens for pale, delicately dotted sanddarter gobies (*Thrichonotis setiger*). They are also home to lookout gobies which partner earth-moving companion shrimps. On the white expanse many cryptic trails start and end mysteriously – evidence of molluscs that bury themselves by day and only emerge at night.

Thousands of gems are hidden in each handful of sand. There are minute molluscs in exquisitely twirled shells, perfectly formed and functional. We can see the faint, bumpy bruises made on the surface of the sand by hidden sand dollars and heart urchins, while at night we may perhaps come across the mermaid's comb (*Murex pecten*) and witness how it skilfully captures its prey by slamming its exquisite tines cagelike over the victim. We may find the beautiful harp shell sliding along ready to drop the tip of its foot as bait.

During the day, large colonies of garden eels (*Heterocongridae*) protrude inquisitively from their burrows. They sway sinuously to make social contact and to feed on plankton soaring by. At the slightest sign of disturbance, they disappear downwards; beady eyes and snout are all that remain to betray them. Only the most patient diver will see them emerge again.

Where small coral patches form islands of shelter, the speckled sandperch (*Parapercis hexopthalma*) lurks quietly, propped up on dainty pectoral fins, staring down divers and prey.

Often, strange heaped coils of excreted sand are plopped over sandy flats, apparently without origin. These are the product of the furtive acorn worm (*Hemichordata*) which lives, unseen, in a U-shaped burrow under the depths of the sand.

Above: *'I dare you to see me' is the delicately patterned message of the peacock flounder* (Bothus mancus).

Previous pages: *Blade-thin and mirrorlike, shrimpfish* (Aeoliscus strigatus) *seem a mere illusion hidden amidst the grassy oceanic meadows.*

Magnificent stingrays rest on the white dimensions, skins peppered with a dense sprinkling of sand. At night, setting off to forage for crustaceans, they leave shallow craters in their wake. The flounder, flat and delicately patterned, lies unseen on the grainy sand bed with an asymmetric grin on its skewed face; only its periscopic eyes move quietly in sinister search of prey. The electric ray (*Narcinidae*) cuddles imperceptibly under a grainy blanket, ready and able at any moment to emit a whopping shock to both hopeful predator and unwary kneeling diver.

Goatfish (*Mullidae*) plough up miniature storms and taste for tiny crabs in the sand with wriggling barbels. Persistent chequered wrasse follow in the cloudy wake, profit-feeding on missed bits of stirred-up food. A pair of Pegasus sea moths (*Eurypegasus draconis*) 'walk' in stately miniature gait, almost invisible against the reflected radiance of the sand. Their extended pectoral fins are gossamer, hued with pastel iridescence. The crocodilefish nestles in ambush. His granule-patterned skin is barely discernible and even the glint of an eye is hidden behind a layer of built-in frilly lace.

Caressed by the tentacles of sand-living anemones, the flamboyant clownfish boisterously arrange communal pecking order. Waiting for night's darkness, the delicate filaments of the tubed stingers and the slim bodies of several species of sand eel are safely retracted into cool, dark dens.

Closer to shore, where dank mangrove forests hem the edges of coastlines or where land embraces water in calm, protected bays, the floor becomes slick, dark and muddy. Yet even here some hardy corals find living space. While they fail to display any splendour, they offer a myriad sheltering hides.

Muddy ooze, silty floors and sand are the ideal places to look for pretty sea pens, a kind of burrowing coral named for its resemblance to old-fashioned quills. Some species are able to retract completely when disturbed, others merely sway gently away from danger. They are favoured as multistorey homes for very small crabs and shrimps, species which are seldom seen elsewhere. Sea pens should not be touched as they have very sharp glassy spicules which can cause extremely painful and irritating wounds.

Many species of mollusc are found only here, thriving on the organic floor. Large volute shells (*Volutidae*) come out at night. Also nocturnal, harp shells (*Harpa* sp.) sometimes sacrifice a bit of their large foot muscle. This initially looks like a distracting defence tactic, but there is another reason. The decoy is also used as a lure: while the predator is thus occupied, the mollusc quietly backs up on the hunter, envelops it in a mucus secretion and then proceeds to eat it.

Nurseries of fry are hidden in the shallows or among wispy grasses and algal mats, safe from the voracious feeders of the reef. Specialized shallow-water species flourish in these quiet areas; they are sturdier than their sand-living counterparts and better burrowers. Many kinds of frill-bodied or translucent nudibranchs abound. Mud-burrowing octopus and their miniature cuttlefish cousins make homes in the silt, frequently matching the bottom with chocolate-brown skins. Espresso-coloured cockatoo waspfish (*Ablabys taenianotus*) are related to the sailfin leaf fish and, like them, imitate swaying weeds around rocks and corals where the floor is rich and dark.

Here the carapaces of decorator crabs are densely covered with living bits of weed and hydroids so that they appear like rather tattered and woolly tarantulas in their quest for camouflage. The larger javelin mantis shrimps thrive in this environment, displaying striking orange and pink spears while they survey the surrounds with their comically stalked golden eyes. Sea stars here are often pale and have a bristly appearance or are otherwise unbelievably brightly coloured. Their tube feet move searchingly, lifting and touching in a never-ending slow-motion march.

Below: *The cockatoo waspfish* (Ablabys taenianotus) *imitates decaying leaves on shallow, muddy reef floors close to shore.*

Many urchins prefer this organic environment, sometimes displaying luminous and electric colours at night. Highly poisonous fire urchins glow like hot embers while others have spines that are banded in juvenile colours, between which they display single, beautifully speckled, translucent bubbles that look like solitary eyes – these are the cloaca that fastidiously lead wastes away from other circulatory apertures.

Where mud mixes and marbles with sand, vast sea meadows grow. Sometimes a lone dugong grazes here, while between the dense waving stalks and blades of this marine grass, in very shallow water, the inimitable sea horses (*Hippocampinae*) find a perfect shelter, colour-complemented to such perfection that only the very trained eye can discern them. The trick is to look for the curl of their prehensile tails. Contrary to popular belief, sea horses do not always cling to grasses in their upright swimming posture but rather lean floppily into the current and surge to emulate the swaying movements of the grass.

The specialized eyes of these charismatic creatures move independently through 360 degrees. When they spy suitable planktonic snacks, they suck in the morsels with dainty pipette-like mouths. Sea horses tend to be monogamous, at least for the courting and reproductive season. Their courtship, the transfer of eggs to the male's brooding pouch and the eventual birth of the juveniles is one of the most breathtaking sights a snorkeller or diver could ever behold.

In such a meadow one night, I came upon the very rare and extremely difficult to detect double-ended pipefish (*Syngnathoides biaculeatus*). These pipefish are so effective at imitating green leaves that only the movement of an eye and a momentary lapse in mimicking posture gave it away. Although they have slightly tactile tails and sea horselike faces, they have rather rigidly held bodies that blend with the broad blades of sea grass.

Below: *Dwarf lionfish* (Dendrochirus brachypterus) *usually occur around small obstructions, and when photographed show a curious 'lightning bolt' in their eyes.*

Above: *Award yourself full marks if you can see the double-ended pipefish* (Syngnathoides biaculeatus) *as found here on a night dive.*

Above left: *These ghost pipefish* (Solenostomus cyanopterus) *convincingly mimic small grass blades.*

Several sea grasses stand as model for the shapes of another amazing family, the seagrass ghost pipefish (*Solenostomus cyanopterus*). These tiny creatures usually occur in pairs, hovering next to or in the plants, pointing their thin noses down to suggest stems. Their bodies imitate the leaves exactly in both shape and colour, and their movements are a most believable duplication of the gentle sway of the plants. When the blades of grass die and turn brown, the colours of the ghost pipefish follow suit. Only very close inspection will reveal them to the diver's eye.

On such a grassy meadow, adjacent to a small, flat patch reef in Mauritius, I have often watched the mimicking dance of Indian Ocean ghost pipefish pairs. There is an abundance of a particular decaying weedy leaf on the reef floors. Brown with random white speckles, these leaves are chosen to be copied. The pipefish adopt the exact colour of the leaf, completing the arrangement with matching speckles, blotches and patches of decay. Again, the sway-and-tumble movements of the ghost pipefish are a perfect duplicate of the leaves in the bottom surge. Were the tiny creatures to stop here, few eyes would pick them out. But, as soon as they see a diver, they exaggerate their miming movements as if to make doubly certain of success. That, of course, is what gives them away. Just a bit too persistent, they tumble, sway and sweep, manoeuvring for greater distance from the threat. Their eyesight is quite amazing and they roll their tiny independent orbs all the time to keep the diver in view.

Provided that cleansing and nourishing currents sweep often and gently through these sandy meadow shallows, some very delicate crinoids seem to thrive in these unusual surroundings, clinging precariously to the soft, silty floor. But in Milne Bay, Papua New Guinea, these crinoids also lend their feathery arms as camouflage to the tiny harlequin ghost pipefish (*Solenostomus paradoxus*) – perhaps this also occurs elsewhere. In the Maldives, the Solomon Islands and even Mauritius, these harlequin ghost pipefish can also be found on gorgonian corals.

Named for their beautiful gemlike colours and the many fine, feathery growths that serve to match the pinnate arms of their hosts, harlequin ghost pipefish hover nose down like all their relatives. They also have the orbed, individually moving eyes and use the same method of feeding as sea horses, vacuuming plankton delicacies neatly from the water. The females have adapted their pelvic fins by partial fusion into pouchlike structures to hold their eggs for, unlike the sea horses,

female ghost pipefish guard the broods. The tiny, much scrawnier males match the females' movements in a delicate coordinated ballet. This kind of ghost pipefish can also be found on horny and gorgonian corals, where they mimic the colours of algal or filamentous growths. The exotic ornamental ghost pipefish prefers to live among less interesting bottom debris. When it is alarmed it flares a miniature transparent, pink-spotted fantail. Yet another bright green species mimics the leaves of the calcium-producing seaweed *Halomeda*.

A further member of this family is the tiny hairy ghost pipefish. It has bright orange, hairy growths and bears, at least for now, the common and very apt name of 'Irish setter'. Predictably, this little fellow always occurs near, or disappears into, a specific orange-brown hairy filamentous alga.

The comical box-shaped longhorned cowfish (*Lactoria fornasini*) live in these meadows, mottled in green and creamy-white hues, changing from dark to light as they cruise over various colours. Here, even pufferfish are striped in muted greens and whites, blending into the scenery. Other boxfish which like to live on the lee side of reefs often stray to this environment. But several other fish live and breed here, and they are all suitably flecked with predominantly green and brown blotchy patterns.

Above: *The hairy ghost pipefish* (Solenostomus *sp.*) *looks like a floating piece of algal 'fluff'.*

Opposite top: *Few divers have seen the breathtaking swim of the blue ribbon eel* (Rhinomuraena quaesita).

Opposite bottom: *Javelin mantis shrimps* (Squillidae *sp.*) *use spearlike front limbs to kill and 'kebab' their prey.*

Shrimpfish (*Aeoliscus strigatus*) like to hide between sea grasses or coral branches. They have elongated, bladelike bodies and move in tightly ordered 'parades', swimming vertically with heads pointing down, functioning as one. Every now and then, the stragglers scythe through the water to rejoin their school. Their partially striped and silver bodies seem so translucent that they almost disappear between the grasses, melding with lines of shadow and light.

Anemones with clownfish live in meadows too and in some parts of the southern Pacific, the panda, or saddleback clownfish (*Amphiprion polymnus*), are prolific. Much more restless than other clownfish, when disturbed at night, they may even dive into the gut-cavity of their host. I have also found some tiny medallions of anemones attached to the grass like drooping sunflower heads, in the centre of which a very tiny clownfish or shrimp looks much like a bumble bee. We have been unable to identify these and do not know how they attach to the plants (are they a type of swimming anemone or have they reproduced by fission from the larger ones on the sea floor below?).

Towards the edges of these meadows where patch corals begin to make an appearance, the rapacious mud-living javelin mantis shrimps (*Harpiosquilla harpax*) are abundant. Not only on the reefs but also here, elegant and spectacularly coloured blue ribbon eels (*Rhinomuraena quaesita*) sometimes occur in small, loose colonies, but more often singly. This small eel undergoes abrupt changes in coloration and sex, the only member of the moray eels to do so. Colour phases may vary from vibrant blue and yellow for the male to bright yellow for the female and black with yellow trim for juveniles. It is shy and secretive and, unique among eels, has flared nasal palps and delicate chin barbels. The elaborate nose flares may serve to lure prey. Usually only the head and upper part of the body protrudes from the burrow. The eel has a peculiar swaying motion while it gapes widely, jerking in between as if perpetually straining to sing the high notes of an operatic aria.

The blue ribbon eel has a disproportionately long length. When swimming, the flat body unfolds into a wide ribbon that not only undulates in graceful curves like the ribbons used in rhythmic gymnastics, but also contracts and expands in smooth, accordionlike waves from head to tail and at an incredible speed.

Exotic, fernlike anemones thrive in sand or mud; they are uninhabited and are more viciously armed than their other relatives. Be extremely careful around them as they can inflict horrible stings, the damage of which may linger for ages and cause skin to blister and scar. When such lesser-known species do not harbour clownfish, they are imparting a subtle hint.

The mud, the deserts and the meadows – although they are not strictly part of the coral reef system, they accommodate many of the more interesting and exotic coral reef species in various intriguing ways.

NIGHT'S FAIRYTALE WORLD

The reef after dark has a different nature. The brilliant light of day and the filtering effect of the water mutes many colours to pastels. At night, the reef bursts into deep colour, while hitherto unseen creatures make their debut. Yet for the diver, dependent on torchlight, only parts of the reef are seen in the medallion of light, intricate pieces of a puzzle that can only be assembled later.

Flashlightfish emit a magical morse code; they are equipped with pulsing luminescent organs beneath their eyes, with which they lure prey and see to feed. So radiant is this chemical luminescence, that the fish has to protect its own eyes against the reflection with movable black smudges.

The gaudy fish of the day have changed into drab, muted pyjamas and rest quietly wherever they have found suitable shelter. The quivering damselfish are finally still, hiding in cracks and crannies to sleep. Parrotfish, that normally busily flit by like bits of broken rainbow, are now tucked into their favourite sleeping places. Some are cocooned inside delicate, transparent mucus balloons, shrouds that perhaps hide their scent or act as warning traps. Now, at last, it is possible to approach these exquisite fish closely to witness the intricacy of their luminous tapestrylike patterns and colours. While sleeping open-eyed, they remain completely oblivious to any careful voyeur.

Below: *At night, coral polyps transform into dainty bridal bouquets in seeming contradiction of their carnivorous appetites.*

Wrasse have burrowed into loose sand. Other fish simply lie, deeply sleep-tranced under rather scanty corals and rocks. The night shift is now on duty. Nocturnal hunters, many startlingly red under torchlight, stare from huge-pupilled eyes. While their vibrant colour filters out into camouflaging smudges of shadow during the day, in the dark waters red appears more deeply dark than black. Soldierfish and cardinals forage busily for food. Protected from sharp daytime eyes, slipper and reef lobsters emerge. Even the shy butterfly-winged dwarf lionfish moves out of its cave, half-scuttling on its pectoral and pelvic fins and displaying vivid red flares adorned with glamorous false eye-spots.

The hermit and reef crabs scurry hurriedly, occasionally meeting, dipping and changing direction politely. Some are house hunting, some are scavenging, others are already feeding messily on night-time snacks. The nocturnal nudibranchs graze, quietly sliding their jewelled bodies forward over bumps and minute upheavals. The shelled molluscs crawl from their hides or burrow out of the sand, spreading intricate mantles and gliding after prey on their muscular tongue-shaped feet.

Lionfish twitch and shake their decorated wings. With a flourish of dotted gossamer tails they rise gracefully off the reef into the water column, eyes alert. They hunt with silent stealth. Eel and octopus are archenemies, each keen to prey on the other. But, for now, they slide sinuously over the reef, ready to suddenly propel at lightning speed towards easier and more unwary prey. Basket stars, by day tightly rolled up and deeply hidden from brilliant sunlight, have now clambered daintily to favourite pinnacles and unfurled their magnificent lacy arms. The mere hint of light triggers an alarm that sends limbs spiralling inward upon themselves.

Delicate bristle worms crawl out on their hosts, displaying sparkling tufts of glassy filaments that belie their potential kiss of fire. Tube anemones stream their long, fragile tentacles in the soft current, grasping elegantly at food on the float. The underwater light attracts and concentrates thousands of crazily gyrating planktonic creatures in its beam. As they are pulled towards the tentacled carnivore, its arms begin a greedy, frenzied dance, grabbing frantically at the morsels and curling them towards its mouth. Soon the unexpected and overwhelming glut confuses the anemone into so many strange contortions, that the flower folds in upon itself, attempting to make sense of the meal.

Above: *If the cocoons are left undisturbed, parrotfish* (Scaridae) *can be closely approached at night and photographed.*

If the reef is a garden by day, it becomes a floral fairytale by night. All the corals extend their breathtakingly fragile and translucent polyps, groping greedily at the serving of rich plankton soup. Previously pore-stippled chalice corals now bear layer upon layer of delicate, blossomlike fuzz. The solitary corals bloom with diverse gelatinous bubbles or reach pointed fingerlike polyps to the food of the night. Big-eyed shrimps clamber around on spun-glass legs. Every surface is alive with creatures that venture forth only under cover of the protective cloak of night.

Small Worlds of the Blue Universe

NUDIBRANCHS: GEMS OF THE OCEAN

If they were not so small, nudibranchs would be the star attraction of the reef. Captivating both in their diversity of form and their use of colour and pattern, these molluscs are among the most beautiful of all marine creatures.

Often referred to as 'sea slugs', they have nothing in common with the odious garden variety we know. Somewhere in the course of natural history, these aquatic snails discarded the armoury of their shells in favour of greater mobility. Divers lovingly call them 'nudies' and they are well worth the effort of a concentrated search. Some are so delicately skinned and translucent that organs like the liver, heart, genital glands and radula can clearly be seen. The orange-dotted *Gymnodoris ceylonica* and some aeolids are examples of this translucence. Most nudies, though, artfully combine vibrant and energetic colours. Once you know where to look for them, you could easily see around five different species on almost every dive.

The name nudibranch (pronounced 'new-dee-brank') means 'bare' or 'naked gill'. In some species these naked gills are reduced to conspicuously coloured, feathery appendages peeping out from a small sack on the back, much like miniature bouquets. These gill-sprays can be retracted, especially when the animals are touched. But other nudies may have different gill protuberances on their bodies. Two pretty antennae, called rhinophores, usually match the colour of the gills and can also be retracted into the nudibranch's body for protection. These, together with some small oral tentacles, serve as tactile sensors and chemical receptors.

Nudibranchs are generally tiny, mostly between 3 and 40mm (0.1 and 1.5in), but there are a few large species. The Spanish dancer (*Hexabranchus sanguineus*) is the largest and most dramatic nudibranch, generally growing to dimensions of 25 to 30cm (10 to 12in) long, but in Papua New Guinea we came across one of these beautiful creatures measuring a gigantic 60cm (24in) in length.

Spanish dancers have vibrantly red bodies with either white or creamy-yellow patterns. Similarly coloured, the imperial shrimp (*Periclimenes imperator*) lives commensally under the 'bouquet' of the Spanish dancer's naked gills, moving to different colour patches for disguise. By day Spanish dancers hide in deep, dark cavities, unlikely to be easily spotted. They are seldom found high up on reef walls, usually occurring closer to the sandy reef floor. At night they are more readily found when they crawl out to feed and mate or spawn. Then their dramatic colour is easily seen in a diver's torchlight.

The Spanish dancer is able to swim and it is its elegant undulations through the water, combined with its exotic coloration, that is reminiscent of the flamenco dancer for which it was named.

Above: *The translucent nudi,* Gymnodoris ceylonica.

Previous pages: *Clownfish occur only in their stinging host anemones.*

Below: *Rhinophores are the nudibranchs sensory organs.*

Left: *Tailgater nudibranchs* (Chromodoris *sp.*) *frequently occur in a 'train' formation.*

Below: *Pyjama nudibranchs (here* Chromodoris elizabethina*) are common on almost all tropical reefs.*

In spite of the brilliant colours that the majority of nudies sport, they are often overlooked by divers who simply do not know of their existence or where to look. Many nudibranchs also escape detection because they are either relatively tiny, nocturnal or blend in with their environment. It is only a matter of training the eye; once you have seen a few, spotting them becomes easier.

Most nudibranchs feed by scraping and tearing at food with an organ called a radula, a ribbon of filelike teeth. They only occasionally dine on delicacies such as eggs and barnacles, and tiny crustaceans are coveted snacks. A certain nudibranch on the Great Barrier Reef ingests coral once to obtain zooxanthellae and is then fed exclusively by these algal cells for the rest of its life, never reverting to coral for food again. Others favour *Turbinaria* and *Goniopora* coral species or feeding grounds of ascidian and tunicate colonies. The vegetarian members dine on various carpeting algae and on plant organisms on the surface of rocks or present in filamentous mats and on rich bottom surfaces such as mud.

But, largely carnivorous, most nudibranchs deliberately seek out and feed on sponges, anemones, hydroids, and hard and soft corals.

The choice of diet is intentional, for what they eat is extremely important not only for sustenance but for protection. From these stinging or poisonous organisms, they ingest algal zooxanthellae, which they keep alive internally to obtain coral-matching camouflage colours. These ruffled marine butterflies also separate elements from their food to manufacture sinister defences, for the nudibranch's choice to go unburdened by heavy shells necessitated the development of potent weapons – toxic and distasteful chemicals – to retaliate against predators.

Those that feed on bryozoans and poisonous or stinging sponges and corals ingest and store the potent defence mechanisms of their food undischarged, thereby rendering themselves extremely dangerous to predators. The sponge eaters are perhaps more subtle, yet no less lethal. They win chemical compounds to provide the wherewithal to manufacture distasteful or toxic chemicals, which they secrete in moments of danger. These secretions can often be clearly seen if the creatures are accidentally touched. Others carry spicules in their tough skins or possess replaceable sacrificial appendages that can be cast off to float away, focusing predator attention elsewhere, much like tail-shedding lizards do. Yet other species manufacture concentrated sulphuric acid.

So, the flaunting of brilliant warning colours by almost all of these jewel-like creatures serves to ensure that they will be perceived as unpalatable and inedible. A fish that samples a nudi only once is not likely to forget the experience. The culprit immediately spits out the nudi snack, associates the bold colour with its noxious properties and pointedly ignores the creature thereafter.

Considering that nudibranchs have an extremely short life span – some as little as a few weeks and only very few longer than a year – they employ a clever form of reproduction. They are hermaphrodites, simultaneously functional as male and female. Whenever they meet, sperm packets are exchanged and stored for future use, thus preventing self-fertilization. The millions of eggs, once fertilized, are deposited in undulating gelatinous lace-ribbons which are twirled and spiralled around and attached to various reef spots, after which the parent usually dies. Often these egg masses are as brilliantly coloured as the parents' bodies, many looking like fragile flowering roses. While the colour of the eggs may serve to warn off enemies, I also believe that the floral-looking masses are probably taken for plant growths – at least some of the time – and so escape the attention of reef carnivores.

Pleurobranchs are much larger, and only distantly related to nudibranchs; they are very different anatomically. Commonly called sea hares (*Aplipsia* sp.) because of the sensory protrusions that look like rabbit's ears, they are living evolutionary proof of the process of shell loss, for they still carry the

Opposite: *Imperial shrimps* (Periclimenes imperator) *adapt their patterns to match the unique environment of their Spanish dancer hosts.*

Below right: *Distasteful poisons protect the rose-shaped egg ribbon of the Spanish dancer.*

Below: *The large, flamboyant Spanish dancer nudibranch* (Hexabranchus sanguineus) *is nocturnal.*

remnants of a thin shell under the concentric skin patterns on their backs. The shell disc protects a single gill and other important organs. The fragility of the remnant shell proves to be insufficient for protection, thus the animal is forced to burrow into sand or mud, emerging only at night.

Pleurobranchs feel much harder than nudibranchs and are almost rubbery. Their colours tend to be predominantly golden-yellow or deep wine-red. Some bear breast-like protrusions, others have delicate floral patterns. These animals duplicate the typical egg-mass ribbons of nudibranchs. Some sea hares escape an attack from predators by ejecting a purple fluid.

Below: *Pleurobranchs still retain a very thin internal shell to protect their most delicate organs.*

After years of observation and research, we are beginning to guess quite accurately what foods nudies eat. If we invert this knowledge, we are able to indicate what anatomical characteristics you could expect when you are looking for nudibranchs in the different feeding environments. A close search is often necessary, however, for many may perfectly match the animal or plant on which they feed.

On stingers such as bryozoans, stinging soft corals, hydroids, and ascidians, nudies are almost always brilliantly coloured, thereby cautioning predators to stay away. Many of these nudibranchs have fingerlike growths, called papillae, the tips of which are colour-accentuated.

On sponges nudies may be smooth, but it is more likely that they will have harder bodies with bumpy ridges or warty surfaces. Some are thick and fleshy and sport only one body colour with contrasting rhinophores and gills and a thin matching line around the edges. Or they will have sponge-imitating surface textures or even cell patterns that mimic those of colonial sponges. Chalky-white spots and irregular blotches are not unusual on these nudies. Many display bright longitudinal stripes and brilliant contrasting colours on the edges that match the gills and rhinophores. These colours warn against the poisonous properties gained from toxic sponges.

A few species of nudibranch feed on molluscs (*Ophistobranchs*), for which they become translucent, adding subtle, semi-opaque spots. But completely contrasting spots are not unusual either and the patterns may be used largely to mimic either the mantle or the foot of molluscs.

Soft-coral-eating nudies have much smoother bodies and they often carry polyp-imitating patterns or tufts. Those that have papillae are more sparsely adorned, but the protrusions are much thicker than on hydroid eaters, and always curved. Look very closely, as those nudies that feed mainly along the trunk of soft corals are finely but irregularly striated to blend with the spicule patterns.

On gorgonians, one of the triton nudies is bright blue with uneven red-rimmed circles. In photographs they are so unusually prominent that one expects to see them straight away, but this is simply not so in natural light. Perhaps the blues look like water and the reds like connecting gorgonian tissue? Most other gorgonian-feeding nudies have much less conspicuous bodies, with spaced, paired lateral tufts that imitate polyps or polyp calices, the tiny pores into which polyps retract. One nudibranch we know of feeds on barnacles, two others on sea pens. On various sea anemones, nudibranch bodies look more like aspic jelly, with just a few highlighting marks.

Once divers begin to 'see' nudibranchs, they find that they suddenly become fine-tuned. These intricate creatures soon seem to be like jewels spilled over the reef and one can simply not imagine how it was possible to miss their glorious colours and patterns before.

LIFE ON THE FEATHER STARS

Feather stars, correctly called crinoids, look like giant upside-down tassels, popped randomly in the blue by a decorator gone mad. Extremely ancient plankton-feeding life forms that thrive in nutrient-rich waters, they are especially abundant in areas that have periodic strong currents. To reach the most advantageous position, they clamber up tall structures on tiny feet, called cirri, and extend their tentacles in elegant fountainlike sprays. They are almost always perched on the highest points of the reef, propped on the rim of large elephant-ear sponges or delicately poised on the outer edges of giant sea fans, whips and wire corals. Here they tenaciously cling on, either with a bunch of curly, rootlike attachment feet called basal tendrils, or by using the prehensile powers of their many arms.

Right: *Far from being only extravagantly decorative, feather stars are veritable forests that hide and feed tiny inhabitants.*

Below: *A tiny filefish* (Pseudomonacanthus *sp.*) *in crinoid arms.*

From these living platforms, they grasp at food particles with slow and elegant sweeping motions of their filigree arms. These arms occur in multiples of five, arranged to rise from a tiny cuplike body. Some feather stars have only five arms, others have over 200. Small lime plates, embedded beneath the skin, strengthen the feather stars and give them their prickly feel. The arms branch into many pinnules, delicate growths that impart the crinoid's feathery look. Both the arms and the pinnules bear suckerless tube feet that contain rich mucus-secreting glands. The mucus coats the arms and pinnules with a sticky food-trapping substance. The food, once snared, is conveyed to the mouth via special gutters along the centre of each arm, helped by the sweeping action of the millions of tiny hairs that line them. The crinoid's mouth is situated on the upper side of the cup, as is the anus which is hygienically elevated on a special cone. Crinoids are considered nocturnal, but they are opportunistic and often feed during the day, especially in deeper waters.

Crinoids can reposition themselves when they need a more nutritious location, but some species move around so much that they are commonly called the 'walking crinoids'. These, particularly the *Antedon* species, also relocate by 'swimming' along with currents, simultaneously raising and

lowering alternate filigree arms in a flapping motion. Having found a more lucrative spot, they gently drift down and once again settle into the familiar feathery spray.

Many crinoids are dotted, striped and streaked in combinations of different colours or amazingly geometric patterns of black and white, while others seem to explode with a single shade of neon yellow, orange, scarlet or burgundy. Although we are not certain, these colours could indicate that crinoids are either distasteful or contain toxins.

Because they are largely ignored by other marine life, crinoids afford spectacular homes for some specialized creatures; these true 'crinoid critters' are never found anywhere else. In the depths of

Above right: *Related to gobies, clingfish* (Discotrema crinophila) *have small abdominal suckers with which to latch onto crinoids.*

Right and below: *For mating convenience, crinoid shrimps (right) and elegant squat lobsters* (Allogalathea elegans) *(below) usually pair permanently in their crinoid homes.*

crinoid hearts or on the intricately feathered arms live elegant squat lobsters (*Allogalathea elegans*) which, despite the name, are actually crabs. Many are bicoloured to match their feather star hosts, but in red feather stars they are a deep velvety-red.

The slender clingfish (*Dicostrema crinophila*) is a relative of the goby and aligns itself with the crinoid's inner arms, adhering to them with a tiny abdominal sucker. Some are striped, others spotted or leopard-patterned. The crinoid brittle stars, which are still largely unidentified, weave through the feather stars and cling haphazardly to any available surface. Several species of crinoid shrimp, such as *Periclimenes tenuii* or the horned *Periclimenes cornutus*, as well as the snapping shrimp, *Sinalpheus stimpsoni*, also spend their lives on crinoids, usually in pairs. These tiny creatures pick small quantities of food from the crinoid's supply, but probably also catch plankton themselves.

Though tiny, crinoid creatures are also masters of camouflage. The little clingfish, for instance, rarely exceeds a length of 3cm (1in). If the crinoid has rolled up its arms in slack current, it is even more difficult to search for them. Sometimes, a very gentle tickle on the basal tendrils will encourage the crinoid to relax and unfurl its arms. Occasionally nudibranchs, flat worms and tiny molluscs may also be seen on the stars.

On reef walls it is not unusual to see small schools of tiny fish fry swimming amongst the crinoid arms. Some fish deposit their eggs under or close to feather stars. Once hatched, the fry can unobtrusively hide and grow safely in a home ignored by other fish.

Feather stars are delicate and brittle, and removing them from their perches is almost impossible for the uninitiated without breaking some arms. In addition, an accidental brush of a diver's wetsuit against these creatures will cause them to cling to the fabric instantly. It is impossible to remove them intact and repeated scrapings result in a slimy mess. Luckily, like other sea stars, they can repair accidental damage by the process of regeneration.

ANEMONES AND THEIR CLOWNFISH

Sea anemones are the hungry flowers of the sea. There are as many as 1000 different kinds spread across the world but relatively few are visible. Most live well hidden in cracks and crevices, under rocks and corals, in the sand or in self-made tubes. Others are minute.

The most obvious and best known sea anemones are those harbouring the colourful commensal anemone clownfish – only 10 species, all of which are found only in warm tropical waters. They occur in the shallows, and only seldom as deep as the maximum of about 50m (164ft). Although these anemones also live in a variety of other, sandier habitats, they are most prolific on reefs and so are considered reef-dwellers. Ranging in size from just a few centimetres to well over a metre (3ft), most are found as individuals; if conditions are perfect, they may occur in huge colonies.

We are normally so fascinated by the individuals that live in these tentacled castles that we take little notice of the interesting host. Anemones are invertebrates, lacking both a skeleton and backbone. They adhere to a hard reef substrate with a suckerlike foot, the pedal disc, which elongates into a fleshy, pliable tube or cylinder, the column. This column is smooth in some species and warty or bumpy in others. The tube widens into an upper crown of tentacles arranged around a mouth. Called the oral disc, this is the most visible part of the animal. The disc can spread wide open, for optimal light exposure, or can fold upon itself in a ball shape, enveloping all the tentacles.

The outer surface of the column and oral disc may be coloured in shades as astonishing as hot-pink, golden-yellow and velvet-red, explaining why anemones are valued as such superb photographic subjects. The tentacles among different species are equally diverse in shape and texture. Some are short, stubby and compact; others translucent and balloon-tipped with delicate opaque striations; yet others are shaggy and carpetlike; or long, stringy and spaghetti-like. Added to this confusion of shapes, the tentacles may also be tipped in pink, purple, blue, red, green, white or brown. These colours are imparted by zooxanthellae and even in the same species may be completely different or even absent. This random and fickle way of nature often stumps us when we want to identify different species.

Below: *Nestled deep in its home, a clownfish* (Amphiprion bicinctus) *temporarily turns 'punk' under its tentacle crown.*

Right: *These beautiful colonial anemones* (Nemanthus annamensis) *settle only on sea fans.*

Below: *A single balloon-tipped tentacle* (Entacmaea quadricolor) *makes an exquisitely translucent macro subject.*

But at least anemone tentacles have one thing in common: they sting! Belonging to the genus *Cnidaria* (the stingers), anemones are related to both sea jellies and stinging corals. The tentacles contain taste cells that indicate if caught prey is edible. They also harbour thousands of microscopic harpoon-like nematocyst cells that can fire to capture and paralyze living food. The tentacles drag the prey into the mouth and stomach, where it is dissolved and digested. The wastes are then disposed of in a reverse process.

As mentioned before, anemones harbour colonies of zooxanthellae, the one-celled algae that, as in corals, need sunlight for photosynthesis. The cells produce sugars from carbon and water, nourishment that they use themselves but which is also 'leaked' to the anemone and may often be its main energy source. Additionally, depending on currents, sea anemones also feed on plankton. Only occasionally do they supplement these sources by feeding on animals such as small fish, crustaceans and urchins.

Some species appear to burrow into sand, but they actually adhere to buried hard substrate. Many of these are capable of retracting into the sand with a fast spiralling motion when they are touched or disturbed by an invader. Of course, the resident clownfish then temporarily lose the protection of the anemone until it deploys again, something which may result in their death or the loss of their eggs. Hungry predators and egg-devouring wrasse are always around!

The adhesive anemone species are very short-tentacled and also able to withdraw. They feel particularly sticky to the touch and, with one exception, do not house clownfish but almost always harbour anemone shrimps.

When anemones settle in a favourite spot they seldom move again, although they are capable of a very slow slide on the pedal disc. While they have a considerable life span, sometimes exceeding 100 years, they sadly have a painfully slow reproductive rate.

The stinging cells of anemones also serve to repel predators. Even so, we may safely touch the tentacles of most anemones as the hardened skin of our fingers or palms cannot be penetrated. However, greater care must be taken to avoid contact with more sensitive skin on the wrists, arms and face. Remember that some anemones sting more than others, and all may sting more virulently depending on the season.

The anemone constitutes the entire self-contained territory of its anemonefish. Because we know that anemones can capture and eat fish, the incredible relationship between this stinging host and its small, defenceless partner is nothing less than pure magic. The 'diplomatic' immunity that these fish enjoy has fascinated scientists for a long time, and even today a conclusive answer has still not been found.

We know that anemones can discharge their nematocysts without harming their resident anemonefish. Any capture of prey by the anemone is activated through a combination of physical contact and a reaction with chemicals present in the prey's mucus. Yet the anemone simply does not recognize the clownfish as food. It is thought that the mucus layer on the clownfish's skin, after a period of acclimatization, either changes or develops special chemical properties that prevent nematocysts from discharging. Whether the chemicals are manufactured by the fish or by the anemone is not known.

It has been proved that a fish, when removed from its anemone for a while, loses its immunity. When the partners are reunited, the fish must perform a repeated

Following pages: *This clownfish* (Amphiprion chrysogaster) *is strictly endemic to the Indian Ocean island of Mauritius.*

Below: *Nocturnal tube anemones* (Cerianthidae) *are frequently inhabited by anemone shrimps.*

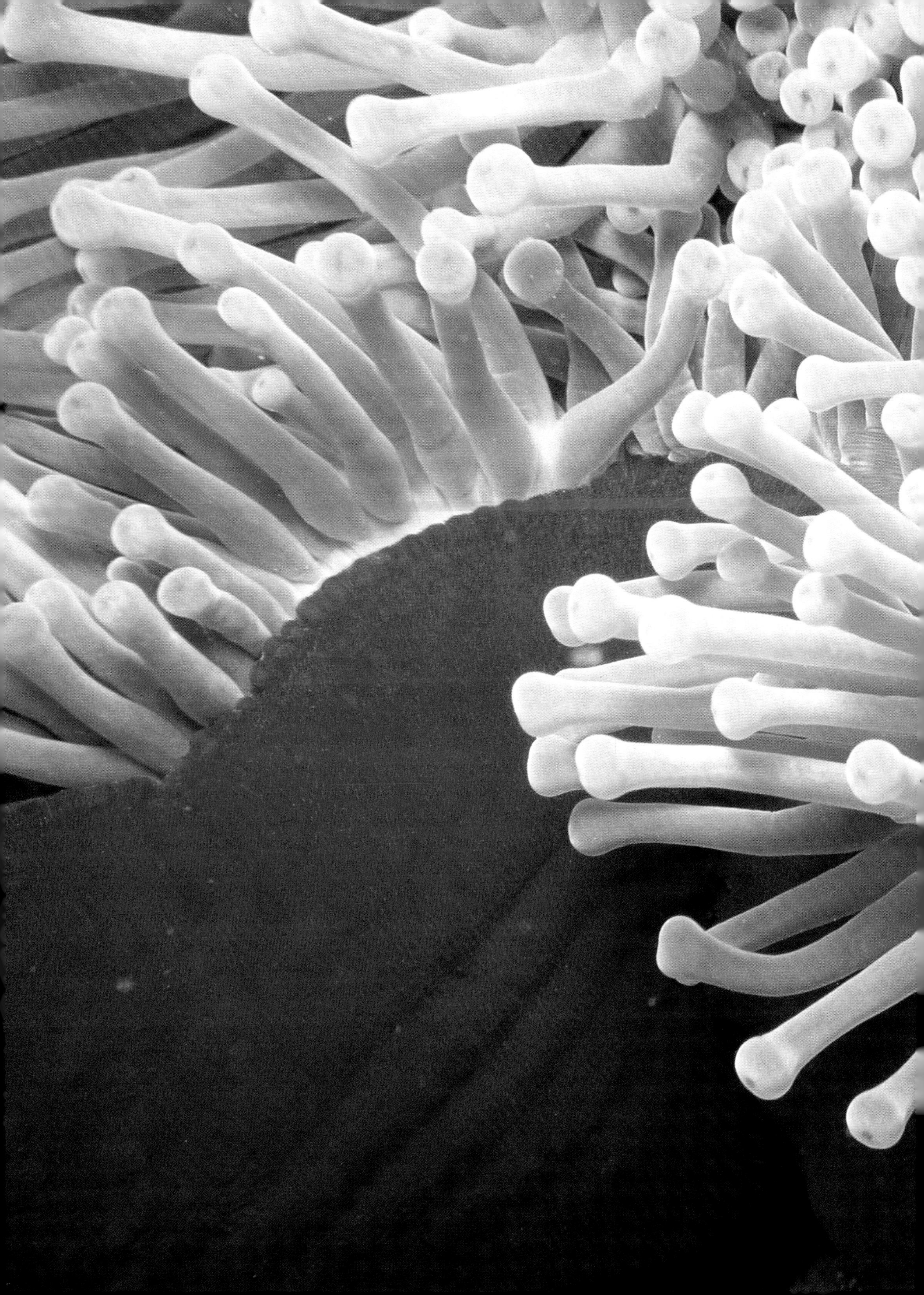

Classifying Clownfish

All 28 species of anemone clownfish are members of the Pomacentridae, *or damselfish family. Twenty-seven carry the genus name* Amphiprion. *But one single species, the breath-taking spinecheek clownfish, is so different that it falls in its own genus, namely* Premnas. *Few clownfish are host-specific and most will readily settle in various anemones, but the tomato clownfish (*Amphiprion frenatus*) and the spinecheek clownfish (*Premnas biaculeatus*) seem to occur only in the anemone* Entacmaea quadricolour, *while the pink anemonefish prefers* Heteractis magnifica.

touch and withdraw acclimatization process. Starting first with only the basal fins, the fish exposes progressively more of its body until it is once again immune. The protection is thus acquired and can be lost. When an immunized fish is placed in the anemone of another anemonefish, it will initially also be stung, until it has once more repeated the immunization process. This behaviour suggests that the fish, rather than the anemone, might induce the protection. However, it is equally likely that a mysterious combination of factors may be at work.

The profit for the fish in this commensal relationship is obvious and simple: protection from predators. The advantage for the anemone is less apparent. The clownfish wards off tentacle-nipping predators such as butterflyfish, but also grooms the anemone by removing debris, thus contributing to its general health. In reef experiments it has been proven that the anemones may disappear shortly after their fish are removed and, although no one knows for sure, there is strong evidence that voracious butterflyfish finish off such anemones in a matter of minutes.

Clownfish are unable to live without their anemone hosts and are therefore never found elsewhere on the reef. In the commensal, or symbiotic, relationship between the two the clownfish is thus the 'obligate' or 'obligative partner'.

However, in a protected aquarium the clownfish can live successfully without anemones. Aquarists should refrain from keeping anemones as they do not survive well in this environment, except where natural planktonic waters are used. They soon begin to droop, looking shabby and stunted in the absence of sunlight and natural currents, until they eventually die. Because of continuous replacements in aquariums, these animals are now over-collected and disappearing at an alarming rate, much too rapidly for natural replacement. The slow reproduction rate and the odds against which any clownfish finds suitable anemones on reefs should be considered. Imagine what contribution aquarists could make if they simply refused to buy these animals.

Clownfish may live alone in an anemone but they normally occur in pairs or small communities. The female is the larger fish and she is dominant in the hierarchy; the male is the smaller adult. The other members of the community appear to be juvenile but they are not necessarily young. Instead, their development is retarded until something happens to an adult or until another anemone becomes vacant. These 'stunted' members are also not the offspring of the adult fish pair (except in aquariums) as Nature has taken pains to ensure suitable dissipation of genes.

Clownfish spawn throughout most of the year, although sexual activity intensifies during the warmest seasons. Unique among damselfish, clownfish pairs form permanent bonds. For several days before actual spawning, social behaviour in the anemone host is acutely heightened. There is much swimming and ritual posturing, fin erection, signal jumping and nest preparation. The male may become very aggressive, chasing and nipping his mate. He selects the nest site adjacent to the anemone and spends considerable time clearing it of algae and debris. If a suitable substrate is not present, clownfish have been known to drag flat objects like stones, coconut shells or small dead coral slabs towards their anemone. Closer to spawning, the male is joined in the preparations by his mate. Eventually, the tiny conical ovipositor of the female becomes visible and spawning begins.

The female then follows a deliberate zigzag pattern over the nest, depositing her eggs while the male follows her path to fertilize them. The eggs are attached to the nest with short tufts of filament, forming the furry reddish or brownish patches we often see next to or slightly under the flap of the anemones. Throughout incubation (six to seven days), the nest is meticulously maintained and zealously guarded by both the male and female. They aggressively chase away potential egg-eaters many times their own size, returning to the nest to fan the eggs and remove any debris.

Despite the fact that clownfish are truly lovable, courageous characters, they are sometimes so protective of their nest that they will attack a territory-invading diver or photographer fiercely

Far left: *The spinecheek clownfish* (Premnas biaculeatus) *is the most flamboyant of all clownfish species.*

Centre: *Clownfish care for and aggressively defend their eggs against any intruders.*

Below: Amphiprion ocellaris *is unfairly called the 'false clownfish' because it is so often confused with* Amphiprion percula.

(believe me, they can nip out sizable chunks with their sharp teeth!). Their agitation at intrusion must be understood. It is kinder to back off or to find another, more accommodating community.

The larvae hatch, rise to the surface and become planktonic for about eight to 12 days. Those that survive return to the reef as small fish and must rapidly find a host. It is possible that this problem is overcome by anemones releasing chemical trails in the water, or that the fish find anemones only by sight – we simply do not know. Even after having found an anemone, the clownfish's troubles are not over. Residents already in place will make life difficult for the newcomer, for there is a definite pecking order in the community, with much 'bullying' from the adult male down to the smallest fish.

The phenomenon of sex reversal is an interesting element of fish life and one present here too. But unlike the usual female-to-male change, it is the male clownfish that undergoes the sex change, usually when the dominant female dies or is otherwise eliminated from the community. When a vacancy occurs, the dominant male will become the dominant female within hours, while the next male in the queue will grow rapidly to adult proportions.

Clownfish display a heady variation of colours and stripes, and all are used for scientific identification. It is therefore important, when trying to identify a species, to note exactly how many stripes are present and whether or not they reach to the very edges of the body; also whether stripes are distinctly lined by black and how wide the black markings are. Sometimes vast colour differences occur within the same species, mostly dependent on age or sex. Such colour differences within a species may also be due to hybridization or may simply be random.

The anemone is the hub of existence for other creatures too. In some hosts, the tiny dominofish – of the genus *Ascyllus* – are the most abundant and obvious. They may sometimes be the only inhabitants, but generally they will share the anemone or at least its proximity with the clownfish. They are facultative symbionts. Most species that are considered to be facultative symbionts seem to avoid direct contact with the tentacles, simply using the protective

ANEMONE HOSTS

The 10 anemone species which host clownfish are:

- *Bulb-tentacled anemone* (Entacmaea quadricolor)
- *Corkscrew anemone* (Macrodactyla doreensis)
- *Magnificent anemone* (Heteractis magnifica)
- *Delicate anemone* (Heteractis malu)
- *Beaded anemone* (Heteractis aurora)
- *Leathery anemone* (Heteractis crispa)
- *Mertens' anemone* (Stichodactyla mertensii)
- *Gigantic anemone* (Stichodactyla gigantea)
- *Haddon's anemone* (Stichodactyla haddoni)
- *Adhesive anemone* (Cryptodendrum adhaesivum)

proximity of the anemone; however wrasse from the *Thallasoma* family infrequently make contact and seem to come to no visible harm.

Anemones also harbour several crustacean species. There are usually paired shrimps, and as many as eight individuals may live on the tentacles or under the flaps of larger anemones. The *Periclimenes* and Amboin shrimps are the predominant species. Many are exquisitely transparent, and when such a female carries eggs, they can be seen clearly through her glasslike body. Spotted porcelain crabs (*Neopetrolisthes* sp.) also live in, on and under the host anemones.

These small shy residents are difficult to effectively mark or tag, so they are not particularly well studied and we still do not know how their lives are tied up with those of the anemones or whether they are obligate or facultative.

Opposite: *Panda clownfish* (Amphiprion polymnus) *share living space with porcelain crabs* (Porcellanidae *sp.*).

Above: *Some clownfish actually bite off tentacle tips and threaten to sting persistent intruders.*

Left: *Whitebonnet clownfish* (Amphiprion leucokranos) *usually cohabit with different clownfish species, presenting riddles about their mating habits.*

HEALTH CLINICS AND GROOMING SALONS

Below: *A cleaner wrasse engages in meticulous dental work for a spotted sweetlips* (Plectorhinchus chaetodontoides).

The marine population at large is vibrant and healthy, and sick or ailing fish are very seldom seen. Naturally, in the scheme of selection, the weak and mortally wounded are removed almost immediately. However, an intriguing and very visible relationship between marine creatures and specialized body cleaners is what contributes most to reef animal health. This relationship is present on almost every reef and may be closely observed once you are able to locate the cleaning stations.

The pests and the parasites of the sea are mainly crustaceans. Fish may partly rid themselves of these annoyances by either using their own fins or rubbing against rocks and sand or even against each other. Jackfish and trevally often avail themselves of a rub against the rough skins of sharks, if these are present. However, the only sure way for them all to solve the problem without coming to personal harm is to utilize the services of the specialized cleaners.

It is known that some juvenile wrasse, angelfish and hogfish perform cleaning services when they are young and defenceless, perhaps for a short while using the strategy of 'not being eaten in exchange for services'. On one occasion I even observed juvenile Moorish idols cleaning damsels, but these fish are not true cleaners.

The true lifetime specialization of cleaning is actively practised by some 25 different fish species from eight families, six shrimps or prawns and at least one crab and one worm. The most specialized and best known cleaners are wrasse from the genus *Labroides* (*bicolor*, *dimitiatus*, *rubrolabiatus* and *phtirophagus*). In the Caribbean, several species of tiny goby fulfil the cleaning role.

The striped lady cleaner shrimp (*Lysmata amboinensis*) and the banded barber shrimp (*Stenopus hispidus*) are the best known and most frequently seen shrimps that specialize in the cleaning process. Other cleaner shrimps are *Rhinochocinetes uritai*, *Stenopus pyrsonotus* and *Stenopus zanzibarius*. But it is the first two shrimps and the little blue-streaked cleaner wrasse that we most often see and classically associate with most cleaning stations.

Opposite: *The painted cleaner shrimp* (Lysmata amboinensis) *fearlessly enters the mouth and gills of a coral grouper* (Cephalopholis *sp.*).

The cleaner wrasse set up professional health-care centres together with their cleaner shrimp colleagues. The location is almost always situated at the higher spots on a reef and is soon known to all inhabitants. From here, the fish advertizes its business hours with a characteristic bouncy dance routine, invitingly flashing its signal streaks of electric blue. The shrimps are found lower down in dark recesses, where they perform a ritual tap dance while simultaneously waving their long, white, contrasting antennae. Together they entice almost every mobile creature around to visit them.

Above: *Moray eels often keep teams of cleaner shrimps* (Lysmata amboinensis) *in their dark lairs.*

When the reef is prolifically populated, residents regularly form orderly queues, patiently awaiting services. Large nonreef, open-water species 'come to town' specially for the full treatment from time to time. Predatory sharks, manta rays and big groupers habitually visit and line up at favourite stations. More nervous species will hover carefully in midwater above or beside the stations, waiting for the cleaner to approach them.

After potential clients have been attracted by the luminous blue flashes, both cleaners repeatedly perform elaborate rituals of approach, pose, touch and retreat. These rituals induce the client to assume a slightly oblique posture, which signals safety and invites commencement of services. The ritual very often comically resembles that of eager attendants cosseting and sales-pitching clients at an exclusive hairdressing or beauty salon.

The cleaning territory also seems to be a safety zone. Although posturing fish always seem to keep a wary eye on the surrounds, they are to my knowledge rarely attacked. As soon as the ministration starts, the client seems to go into a kind of trance. Fins are spread, gill covers lifted and mouths are kept wide open. The tiny cleaners do not hesitate at the mouth of fish many times their own size and the client fish never makes a mistake. The wrasse and shrimps respectively use their tiny teeth or pincers like tweezers to remove ecto-parasites, loose and dead skin, and mucus – all ingredients upon which they feed. This is the payment for the services. At times these incisions must hurt but, even though the clients might wince or shake the cleaner off for a moment, they generally endure the necessary surgery with patience. Wounded and infected fish make extensive use of cleaner services. They consequently heal very rapidly, almost always without scars. This is why cleaners are also called the 'doctors of the reef'.

This splendid arrangement is sometimes exploited by a devious impostor, the sabretooth blennie. It has blue stripes only slightly different to those of the cleaner wrasse. Thus, advertizing under the guise of its model, the opportunistic 'false cleaner' simulates the dance, courts ignorant clients and is unwittingly accepted. But unlike the true cleaner, it has an underslung mouth and large, venomous teeth. It slips in for a lightning-quick bite, removing a semicircular chunk from the horrified victim. When the alarm is given, all fish instantly tense and avoid being cleaned, and the genuine cleaner will, at least for a while, be very closely inspected or avoided entirely.

But fish are astute – creatures learn from a young age what is harmful or dangerous to them. So the impostor usually concentrates on younger, more naive fish. That is why I am not convinced that the disguise is used for predation on a regular basis. To me it is more plausible that the fish employs the mimicking tactics mainly as pretence, to signal innocent usefulness and to avoid being predated upon. It may only make occasional use of the 'cleaning' opportunity, presented on such an attractive silver plate, to supplement its diet.

The shrimps restrict their operations to the dimmer areas of the station. The adult pairs are monogamous and may remain at the same place for years. They require their clients to relax either under a small overhang, sheltered between rocks or in holes and caves. There the shrimps enter the mouth and gills of the client, often climbing in one opening and exiting from another. In general, I suspect that the wrasse attend to the main 'body work', leaving delicate, detail work to the shrimps. At night the shrimps continue, cleaning even sleeping fish. On one night dive, I came upon a sea cucumber and an octopus sitting side by side, being diligently examined and cleaned by several adult pairs of shrimps.

Groupers, and especially eels, habitually lure personal attendant wrasse or pairs of shrimps to their private lairs; both the cleaner wrasse and shrimps seem very enthusiastic about cleaning eels and settle permanently in an amicable 'master and valet' relationship. In the Indian Ocean, large moray eels normally have shrimps in their dens. The blue-and-yellow cleaner wrasse (*Labroides bicolor*) seem to be at the exclusive disposal of eels in these waters, often in pairs, leading a nomadic lifestyle as they follow their 'masters' wherever they go.

It is only when I saw a cleaner wrasse attending to a clownfish in its anemone that I realized there are roving cleaners that make 'house calls'. Naturally some reef inhabitants are

Above: *A masked pufferfish* (Arothron nigropunctatus) *enjoys the attention of a cleaner wrasse.*

Below: *If divers 'signal' correctly, cleaner shrimps* (Lysmata amboinensis) *will afford them cleaning services!*

unwilling or unable to move to the stations. Anemone clownfish belong to this group. Fish that guard a brood will not leave their nest for many days either, so they are also regularly visited and cleaned. Both batfish and puffer-fish are very wary and prefer to be cleaned in midwater. Ever-vigilant, some cleaners have grabbed the market gap! Although they belong to the same species that operates permanently at stations, they seem to have opted for a travelling lifestyle and are not attached to a specific cleaning station.

Even a diver can enjoy being cleaned by shrimps if care is taken in the approach. I have found that shrimps soon hop aboard hands if an open 'handshake' position, thumb pointing up, is presented. Cleaner wrasse are more difficult to fool, but they may inspect and nip at air bubbles captured on tiny body hairs, provided one dives without a wetsuit.

Right: *Well-known cleaning stations are visited regularly, even by creatures as large as manta rays* (Manta birostris).

Below: *These surgeonfish* (Acanthurus *sp.*) *can instantly change colour when cosseted by cleaner wrasse* (Labroides dimidiatus).

Fish display various responses while being cleaned. Some assume a face-down position, others face upwards, some may rest on the bottom or under ledges, rolling from side to side in ecstasy. Many may subtly or drastically change colour from dark to pale; others like goatfish turn rosy during the process! One of the most enjoyable events I ever witnessed was that of black surgeonfish being cleaned on several Papua New Guinea reefs. Sometimes a particular surgeonfish being cleaned would suddenly turn totally grey-white, while its beautiful geometric scrawl pattern became more pronounced. As soon as cleaning ended, the fish immediately returned to its normal black colour and swam off.

The importance of these tiny workers on the world's tropical reefs may be much larger than we think. In experiments on small test reefs, all cleaners were removed, resulting in the desertion of all the other reef fish. The presence of the cleaners must therefore be vital to the community. This should be a warning to all of us, as the cleaner fish are particularly sensitive to water conditions. Water pollution might very well strike at them first, resulting in the bleak and colourless reefs now so evident around heavily populated and polluted resorts.

Cephalopods: The Princes of Illusion

The word 'cephalopod' literally means 'head-foot', and it perhaps most aptly describes the octopus rather than the cuttlefish, squid and nautilus, who are also members of this marine family. To me they are the most fascinating of all invertebrates, not only because of their very obvious intelligence but specifically because of their ability to reason and to learn from past experiences.

Cephalopods are close relatives of the molluscs and, like them, have a radula, a mantle that protects internal organs, and a foot that has developed into several arms and tentacles surrounding their mouth. The cephalopods have the most advanced nervous systems of all invertebrates and very keen eyesight that functions remarkably like our own.

As previously mentioned, cephalopods are the only molluscs that can change colour and the only animals that control the colour changes instantly by muscular action. The colour is displayed when pigment-containing sacs are expanded and disappears when the sacs are contracted.

Below: *Octopus* (Octopodidae) *are magicians, able to match instantly reef textures and colours.*

Octopus

The octopus (*Octopodidae*) is perhaps the most entertaining inhabitant of the reef and simply brims with personality. Octopus have eight arms, the underside of which are covered with sucker discs which can function together or individually as needed. Most octopus build typical houses in reef rubble, piling bits of coral and rock around the entrance, but some mud-living species make underground burrows. Many of the burrow entrances accommodate a sentinel urchin, which probably helps to deter intruders. Frequently the burrow will have a secret back door through which a threatened or disturbed octopus can escape.

Octopus are very territorial creatures and always return to their homes unless they are repeatedly disturbed. They are nocturnal by preference but may be seen during the day if a site is relatively undisturbed or if they have become used to the presence of divers on the reef. Their parrotlike beak can give a nasty bite; however, the

EEL VS. OCTOPUS

We once witnessed the attack of an eel on an octopus. It was a shocking experience. An eel, of which we were totally unaware, shot out from its den with a speed that defies description. As it snapped, the octopus started writhing for a foothold while yielding just enough to prevent being torn to pieces or swallowed whole. The eel countered by creating a knot in its body; it looped its tail and inserted the point, then wriggled the knot upwards to the centre of its body in one smooth move. It now had a measure of stable resistance against which it could strain for a better rip-and-tear attack. While the eel repeatedly tore at some tentacles, the octopus struggled valiantly for survival. It was clinging to a rock with a stretched but still free tentacle, while slinging another over the eel's head. Finally it managed to cover the eel's eyes, where it hung on with determination. Effectively blinded, the eel tapped around uselessly, its sense of smell disturbed by emissions from the octopus and organisms in the sandstorm created by the struggle. It let go for just a moment. That was enough for the octopus to squirt its smoke screen of 'ink' and escape. The fight was a short, brutal burst of violence, over within seconds. The octopus had lost two of its tentacles, which would regenerate, but had survived.

deliberate use of defensive armaments against humans is virtually always the direct result of provocation. And, a word of warning: the bite of a normal reef octopus may become very septic and extremely painful from chemicals present in the saliva.

Some octopus are venomous. Ironically, one of the smallest is also one of the deadliest. The blueringed octopus of Australian waters is capable of killing a human with one bite and should never be handled. It is barely the size of a hand span, brown with yellow bands, and iridescent violet and blue designs which enlarge and pulsate luminously during periods of distress and excitement.

Octopus hunt mostly in dim light and are amazing to watch on dusk and dawn dives as they ooze over the reef, then pounce and drop their tentacles over prey. The skin connecting the tentacles looks convincingly like a trapping parachute; prey is felt for and bitten, then delicately snacked on. Many empty gastropod shells show a tiny hole, the result of an octopus which has drilled near the whorl in order to poison its prey, after which it is extracted and eaten.

Divers should always be alert for unusual behaviour while observing these fascinating creatures. When out on the reef, octopus have an elaborate mating ritual with much display of quickened breathing, passing colour waves and voluptuous caressing, before or during which the female suddenly flares and turns pure white while the male flushes very dark. Should you ever see this colour display, stay put and you may find a touching romance unfolding before your eyes.

During reproduction, the male octopus uses a modified arm to transfer sperm, contained in packets, to the female. Very often, when reefs are dangerous, the female may wisely choose to remain inside her 'chamber', letting the lusty male stay outside to take all the risks. While just his sperm-bearing tentacle reaches in to her, the male will be relatively defenceless, tense and uncomfortable. Yet he is always willing to take the risk. The female may store, use or discard the sperm, depending on how she judges her chances of finding a possibly more desirable beau. Finally she lays

clusters of eggs which she suspends from a previously cleaned surface in her burrow. She guards and maintains these lovingly, never eating or leaving her den until the miniature octopus hatch. The tiny juveniles then disperse to find suitable homes on the reef. These offspring, which in their entirety would fit on a fingernail, already possess and use all the camouflage tricks of their parents. In some species, the females die when their babies hatch and some males die shortly after mating.

The life span of an octopus is short, between one and three years. Its main enemies are large fish, eels, sharks, and dolphins. Defending itself against these hunters, it will use any of its disappearing acts: oozing into crevices, melting over the reef and even burying itself in the sand. As a last resort octopus, like the other cephalopods, 'ink' to escape, but they use this strategy in a special way. Directly before doing so, the octopus turns very dark, then expels the dark fluid and almost simultaneously turns pale. This effectively confuses the predator, the dark fluid holding its attention long enough for the octopus to disappear, either by jetting away or by melding with the surface it flees to. Once in its den, it is almost impossible to attack. The octopus will produce a strong jet of water from its siphon, giving the hardiest predator a mighty scare, or roll its arms in such a way that it plugs the entrance, the tough suckered tentacles displayed towards the outside, at times even holding stones.

The impressive intelligence of octopus allows them to have distinctly individual personalities. When they are in their homes it is difficult to lure them out, but one can easily indulge in a bit of gentle hand-holding by offering ungloved fingers. A 'pet' octopus on one of our favourite dive sites often accompanied Danja, perching on her head or shoulders. Every now and then, it would embark on a tactile examination of her face, her mask and the inside of her ears. As soon as its curiosity was satisfied, it would glide off, insert a slinky arm into the pocket of her buoyancy jacket and daintily lift the fishy snack it knew she carried with her. It was shown this hiding place only once!

Although it is unlikely that any octopus will voluntarily attach itself to your body, it is important never to use force to rip off its sucking tentacles. If this should happen, you will not only alarm and distress the animal, but you will also sport some impressive blue bruises for quite a while. If you simply relax and wait, the octopus will release all suction and escape.

Above: *Incredibly intelligent and curious, octopus will literally follow divers if they are 'ignored'.*

Opposite: *Octopus will usually carry on their natural routine around calm divers.*

Below: *The rare flamboyant cuttlefish* (Metasepia pfefferi) *is an exotic species found only in Papua New Guinea waters.*

Bottom: *Cuttlefish convincingly match their surroundings in both colour and poise.*

CUTTLEFISH

The beguiling reef cuttlefish (*Sepiidae*) has a well-developed brain and an extremely sophisticated eye that can perceive its surroundings with uncanny precision. It too possesses a veritable bag of tricks. Cuttlefish skin is covered in a microscopic, dense dappling and stippling of pigmented cells called chromatophores. These elastic, expandable cells contain various colours and are controlled by muscles connected to a nervous system under the management of an exceptionally astute brain.

The contraction and expansion of the chromatophores provides many different and complex colour and pattern combinations. Only a fraction of an instant is necessary for the cuttlefish to match its background with absolute precision. Some colour cells are used all the time for general camouflage, simultaneously expanding and contracting. Others are used fleetingly, mainly in fear, anger or excitement when capturing prey. But there is also another type of cell present in the skin that is able to reflect light. This creates the iridescence sometimes seen in cuttlefish. It is most entertaining to watch these animals adjust and camouflage as they move over sand or reef, light or dark.

Their colour-changing ability is further enhanced by the even more amazing trick of mimicking shapes, postures and stances. Cuttlefish are able to arrange their tentacles with artistic precision into a convincing imitation of branching coral. In conjunction with colour adaptations, they can raise lacy, pointed and puckered appendages on their skin to copy all natural textures, be they of corals, polyp patterns or marine weeds. The cuttlefish is so successful that it becomes truly indistinguishable from its surroundings.

Unlike octopus, cuttlefish have 10 tentacles or oral arms. Eight encircle the mouth and two are prehensile, tucked away in cheek folds. These can only be seen during the brief time when they are capturing food. Two short arms may lift and arch in front of the face; when this happens, it is a sign of alertness or having spotted food.

Cuttlefish, like squid, have diaphanous skirt-like lateral fins, which are really extensions of their mantle. These skirts are used for stability, steering and slow forward propulsion. In the presence of calm divers, cuttlefish are quite unafraid, even curious. They will hover quietly and stare intently. If you display patience and avoid the reaction-seeking poking we so often witness among divers, they will eventually carry on hunting.

The cuttlefish always hunts with stealth. It will only eat living prey and it is particularly fond of crabs. It typically remains motionless, mentally calculating the striking distance. At times it uses a mesmerizing strategy, rhythmically scrolling flashes of colour across its body. These uniform bands of colour pulsate towards the front of the face so captivatingly that the victim is lulled into stupor, at least for that fraction of time necessary for a strike. It suddenly shoots out the long tentacles from its cheek pouches and grabs the prey. Immediately jetting forward, it enmeshes the prey in its tentacles, then bites and tears it to pieces with its parrotlike beak, while its radula acts like a rasp. All of these complex movements happen at an unbelievable speed and feeding takes but a few seconds.

Inside the body, within the mantle cavity, there is a flat, rigid bone that ensures buoyancy. This is the cuttle. It is commonly found washed ashore on beaches or sold as cuttlebone in pet shops. Cuttlefish can also, like octopus, emit the viscous, inky fluid that is used as a smoke screen in moments of distress or for escape. The resultant visual confusion to predators is probably further enhanced by scent-distracting chemicals in the 'ink'. Cuttlefish use jet propulsion to move forwards and backwards and do so with equal ease and at enormous speeds.

During their courting displays, it is the male that becomes quite pale, unlike the octopus. The pair will exchange continuous caresses and there is much waving of iridescent skirts. The act culminates in a graceful copulating dance, after which the male guards the female while the eggs are deposited, mostly at the base of branching corals. The female cleans them with repeated puffs from her siphon while the male defends mother and brood.

Above: *A detail of the sophisticated cuttlefish eye.*

Below: *Once the egg cases of the squid are planted, they are abandoned.*

SQUID

The shy squid is similar to the cuttlefish but more elongated and, instead of the cuttlebone, it has a gelatinous rod called a gladius. Although its chromatophores are less widespread, the squid has a greater variety of iridescent cells. Diving around a light hung over the side of a boat at night enables you to watch closely the incomparable brilliance of their eyes and see how the iridescent mantle-skirts become opaline.

When possible, squid feed voraciously. During periods of occasional opulence, they will hang in or under fish shoals, holding the next victim in their tentacles while still devouring the first. They are, however, fastidious eaters. Refusing to eat the heads and viscera of their fish prey, they remove these and deftly scrape the body free of all scales, preparing delicious fillets of fish!

Males engage in highly ritualized combat, vying insistently for the attention of females by displaying spectacular colours and patterns. Finally, the performance culminates

Above: *The iridescent beauty of squid* (Sepiidae) *is best seen at night.*

Opposite top: *The nautilus* (Nautilus pompilius).

Opposite bottom left: *A detail of the eye of the nautilus.*

Opposite bottom right: *Nautilus regulate pressure internally and can safely ascend to the shallows to feed.*

in a moving mating dance of sheer synchronized extravagance. Afterwards, the opalescent female securely anchors the egg cases, about as long as her body and prepacked with around 200 eggs, in deep crevasses or between the branches of staghorn corals. Then she dies.

Spawning is frequently a mass affair, where thousands congregate to produce their egg cases. While these numbers draw predators like rays and sharks, it is the consequent death of the squids that results in a scavenging free-for-all frenzy. Sacrificing their own lives may well be the squid's greatest contribution to protecting their future progeny from predation.

ARGONAUT

The paper nautilus (*Argonautidae*) is a pelagic, open-water relation of the octopus. Its translucent shell is in fact not a shell at all, but an egg case that is produced only by the female of the species.

From birth, the female secretes a limy substance from special membranes on her dorsal arms, and fashions it into the exquisitely shaped paper-thin case. She clasps the case to her body, securely cradling it in two arms, using it as a capsule for her many eggs. The exterior of the egg case is covered by the secretion membranes as long as the animal is alive.

Towards the end of her limited life span, the female seeks fertilization from the free-swimming male and then drifts towards the shallows. When she finds sea grass meadows and sandy flats, she abandons the case with her thousands of eggs neatly packed into the protected rear. Her purpose fulfilled, she dies. The baby argonauts hatch at intervals and somewhat later move back to the ocean to start the mysterious cycle again.

NAUTILUS

The pearly or chambered nautilus (*Nautilus pompilius*) is perhaps the cephalopod divers most desire to see. This exotic animal lives protected in a beautiful spiralled shell, considered by aesthetes, mathematicians and artists to be the most perfect shape. The intensively colour-patterned juvenile shell gradually becomes paler, with larger white areas indicating adulthood. The nautilus is an ancient creature. To molluscs it is the primordial ancestor, what the coelacanth is to fish. The most primitive and oldest cephalopod, its fossils date back 400 million years and show it to be one of the earlist forms of life in the sea. Today, only four species remain.

The nautilus is a slow and at times even clumsy mover. It swims by means of jet propulsion, always suspended the same way up, straining – often backwards – against the surrounding waters. The siphon can turn laterally to help change its direction. Primarily intended to provide buoyancy, the coiled shell contains internal chambers filled with a gas that can be controlled and adjusted to regulate flotation. The animal thus functions much like a submarine or, for that matter, a diver's buoyancy control jacket.

While it grows, the animal always occupies the last large chamber, creating new chambers when necessary and leaving the old one behind in the spiral. These vacant chambers help balance the animal's increasing weight. The nautilus has numerous tentacles, some tactile, some prehensile and others used specifically for feeding. The nautilus has suction flaps, unlike the suckers of its relative and chief predator, the octopus, and also lacks the ink sac. Positioned on stalks, its large eyes are endowed with great mobility.

From the depths where it lives, the nautilus undertakes nightly vertical migrations to the shallows where it feeds on the planktonic upwell. It also scavenges on dead animals on the sea floor and readily takes bait in traps. When pearly nautilus die, their shells float to the surface and are washed ashore.

Sea Stars: Constellations of the Ocean

Previous pages: *A male cuttlefish displays vibrant colour during courtship.*

Right: *The commensal shrimp* (Periclimenes soror) *is frequently found on sea stars (here* Linckia laevigata).

Opposite top: *Linckia sea stars are a favourite food for harlequin shrimps* (Hymenocera picta).

Below: *When damaged, sea stars can regenerate an entirely new body from just one arm.*

Sea stars (*Ophidiasteridiae*) are commonly, if illogically, called starfish. Echinoderms, they are related to urchins and feather stars. They are radially symmetric and found in many shapes and colours. Some look like fat pincushions, others sport enough arms to emulate a stylized sun. Most stars can be safely handled, except for the poisonous crown-of-thorns.

While the stars themselves are diverse and interesting creatures, many serve as hosts to a myriad shrimps, crabs, parasitic snails, bristle worms and tiny shells. Some single-valved molluscs also live parasitically on stars, where they cling to the surface of an arm. The large cushion star is sometimes inhabited by a strange transparent fish that lives in its gut cavity, or more frequently by the approximately 1cm-long (0.25in) commensal shrimp *Periclimenes soror*.

All sea stars have bodies that are divided into segments of five or multiples of five and each segment contains a full set of internal organs. This biological strategy means that sea stars can regenerate completely from just a small fragment or else may simply regrow a lost limb. This regenerative process is not at all unusual; we often see it on reefs and in tide pools when we find sea stars with an abnormally small limb.

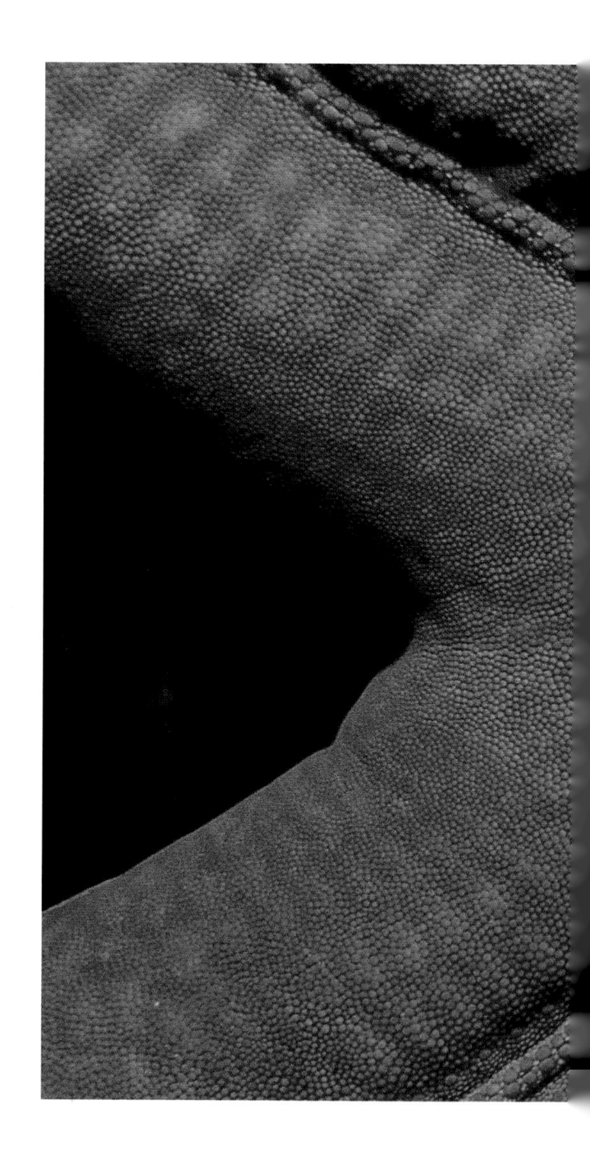

The mouth of a sea star is centrally situated on the underside, which is called the oral side. The anal opening is on the upper aboral side. This indicates that the creature is either bottom-feeding or must clamber upon prey. While many are detritus feeders – the 'sweepers' of the reef – others are carnivorous and feed on sponges, bryozoans, ascidians, corals, urchins, molluscs, and occasionally even on fish.

Some stars feed in a most amazing way! They evert, or extrude, their stomach to surround the edible parts of their prey for digestion, and draw it back once the meal is over. The notorious crown-of-thorns (*Acanthaster planci*), which has devastated huge sections of the Great Barrier Reef, employs this method to devour coral polyps. Originally, when crown-of-thorns stars occurred in pestilential proportions, scientists blamed the over-collection of the giant triton shell (*Charonia tritonis*). This mollusc is a natural enemy of the star and has indeed been removed from reefs in great quantities. But in retrospect the theory is changing to include the possibility that these outbreaks may perhaps be a naturally induced pruning process; these stars feed voraciously on fast-growing corals and in particular on those species that compete for space with the slower but more enduring corals.

After such mass attacks, the crown-of-thorns stars usually and suddenly disappear. While the initial damage certainly looks terribly disheartening, much later many of the reefs are found to have regenerated vigorously. The reason is that not all the coral polyps are eaten by the star – coral tips are left undamaged and able to spawn again. Considering the slow but deliberate scheme of Nature, it is perhaps wise to exercise caution in addressing natural disasters to avoid introducing solutions that may later cause even greater imbalances. This case is still being studied and may yet reveal other surprising facets.

The crown-of-thorns star is dangerous to divers and must not be touched. Its sharp, sturdy spines, covered with toxic mucus, can cause severe injury. Some shrimps adapt to live on this star – the most common is *Periclimenes soror* – but other shrimps placed on it show immediate distress and flee.

All sea stars possess tube feet, some with suction cups, that fulfil functions of locomotion, sensory perception and respiration. The adhesive strength of the tube feet with suction cups is so enormous that the two halves of a bivalve can be prised apart and held open while the starfish feeds.

Brittle stars, sometimes called serpent stars (*Ophioroidae*), as well as the snake stars that wind around the branches of black coral bushes, all lead cryptic lives. They are different from other stars in that they have a central body disc from which the bristly arms radiate. These extremely fragile arms move with fast, winding motions, and are often dropped as a wriggling sacrifice while fending off attacking predators. Even reproduction is possible by detaching one of these tiny arms, which will literally walk away, powered by little tube feet. Called a comet, this piece will generate an entirely new body. Brittle stars are much faster than other sea stars. About 95 per cent are bottom-dwellers which seek refuge under stones and in holes or deep crevices. They are frequently associated with other invertebrates like sponges, which they covet for homes, as well as bryozoans, algae, soft coral, echinoderms and various stinging organisms. However, they may even live in shells, sand and mud. In deeper water, hiding is less evident, probably because it is darker and there are fewer hungry mouths. Few of these brittle, easily breakable stars have yet been adequately studied.

Opposite: *Tube sponges are coveted daytime refuges for brittle stars* (Ophiothrix *sp.*).

Below: *At night, brittle stars (here* Ophiothrix purpurea*) clamber up on reef structures to find food.*

Nocturnal basket stars avoid daylight and remain curled in tight balls, hidden deep within caves and crevices. At night they nimbly clamber to promontories, unfurl their lacelike arms and begin to slowly twist and writhe to snare food in their beautifully poised traps. They are extremely sensitive to light and the slightest flash of an underwater torch will send their tentacles into curling spasms, after which they scurry away to hide, making them almost impossible to photograph well. For best results, photographers must revert to the prefocus-and-lock method with these shy creatures. Left undisturbed, they may use the same hideout and filter-feeding location for many years.

We never pass sea stars without a closer look, for they, together with their various lodging shrimps and molluscs, make great macro-photography subjects. One of the most colourful and exquisite, the secretive painted harlequin shrimp (*Hymenocera picta*), feeds only on sea stars. The shrimp's exotic blotches break up its outline and it is not easy to see. Usually a pair of shrimps will turn a suitable star on its back and pierce the skin with needlelike front nippers. The action invokes immediate swelling, and eases the removal of pieces of flesh. Strangely, the piercing does not seem to disturb the starfish. The shrimps then feed at leisure, starting at the tip and moving towards the centre, before attacking the next arm. Thus, the star is kept alive and fresh until the last arm becomes too small to be mounted. The sea star is deserted and left to regenerate, perhaps as future food for the shrimp. The flamboyant, multispotted flared antennae of the shrimp and the big blades of the third pair of legs are, as far as we know, only decorative. Due to its secretive lifestyle, studies are still incomplete and little is known about its habits. The best chance to see these beautiful shrimps is to find the correct star somewhere in the constellations of the sea.

WEIRD AND WONDERFUL WORMS OF THE SEA

Below: *Christmas-tree worms* (Spirobranchus giganteus) *sometimes occur in huge multicoloured colonies.*

Reefs are well and truly riddled with marine worms. As they are keenly hunted by many reef creatures, they naturally hide from predators and are not very evident. Yet, if we consider that a single coral head may contain as many as 100 species of worm, together with an assortment of other organisms, we can only be astonished at how complex living arrangements are on the reef.

Despite most worms being inconspicuous in colour and habit, Nature acts with typically random artistic abandon: some marine worms are breathtakingly beautiful! The annelid polychaete worms represent around 9000 ringed segmented worms alone. These worms have eyes and chemical receptors and are often decorated with beautiful feathery tentacles. The tentacles are what most frequently attract us. We begin to search for the tube builders, knowing that while these constructions protect them, it also means that they cannot leave their homes. Since they do not have jaws, they must feed by filtering and straining delicacies from the water, which they do with those exquisite hair-lined, feathery bouquets that scientists call radioles.

The most daintily decorated tube worms, and the most often seen, are the multicoloured bottlebrush, or Christmas-tree worms (*Spirobranchus giganteus*). They build hard calcium tubes inside rock or coral and extend a pair of twin whorls – the feeding snares and gills of one worm, not two. The worm can retract these spiral filaments instantly for protection, so frustrating underwater photographers. Additionally, they plug the only entrance to their tube by closing a neat little door, or operculum, behind them, leaving a proportionately large, sharp spine behind to deter hungrily probing predators very effectively. The most lucrative coral head sites are those situated in current paths. They are sometimes so profusely occupied by these colourful worms that they look like festive carnival sites.

The softer and more fragile featherduster worms (*Sabellastarte magnifica*) are yet another example of the beauty that embellishes tropical reefs. These worms live permanently trapped in their hollowed-out tube galleries, from which only their bright, multicoloured, feathery feeding plumes protrude. These exquisite branchiae may be banded in brown, tan, white, purple, yellow or several shades of red.

Above: *The plumes of feather worms* (Protula magnifica) *serve as trapping nets for microscopic food.*

The tiny hairs on the crown of tentacles also help to maximize snaring and directing the food to the mouth. Caught particles are sorted for edibility or for their usefulness in home construction. The organic morsels are ingested while trapped inorganic material is selected, inspected for at least one flat side (to facilitate the building of overlapping tiers), and stored in a special mouth pouch from where it can be retrieved for building or renovating the soft, parchment-like tube. Like all other plumed worms, featherdusters are sensitive to any change in pressure or light and they will instantly retract when approached too closely. If you wait patiently, they will reach out slowly after a while and then suddenly unfurl the living fountain spray. But be careful and gentle, as a second scare will not soon be followed by another show.

The bristled polychaete worms can move about freely because they possess defensive irritant bristles. *Hermodice carunculata*, *Chloeia viridis* and *Eyrythoe complanata* are the three species you are most likely to find. Bristle worms are better known as 'fire worms' and whoever has had contact with these fellows will understand just why. They are really very beautiful, with vivid orange or green bodies adorned with tufts of white bristles along their parapods. But the brilliant colours signal danger, as the bristles can comfortably deter any attack. Normally, the bristles are sheathed and, left undisturbed, the worms are not dangerous at all. But when touched, all the bristles erect, covering the whole worm. Thin, glassy and hollow, these bristles can penetrate flesh and then break off, causing an immediate and very painful burning sensation that may last for several hours. For divers, the only effective remedy is removal with adhesive tape as tweezers only break the brittle shafts off above the skin, leaving the venomous part in place.

The polychaete worms favour coral, stingers like fire coral (a hydroid), and anemones for food. But they are scavengers too and will feed on any dead animals. It is quite fascinating to watch the process for, instead of moving the food to their mouths, they move their bodies onto and over the food, 'engulfing' the meal, which is quickly dissolved by digestive juices.

These worms spawn according to lunar cycles, usually at the same time and as a grand finale to the week-long spawning of corals. The best example is that of the palolo worms, well known in Samoa and Fiji. Prior to the spawning, they develop a special posterior 'tail' in which their sex cells are stored. Then at night, about a week after the full moon in October and November, at the correct time and synchronic with all the others, they spawn. The sex part, or epitoke, disengages and rises to the surface, where it wriggles until all the eggs and sperm are ejected. Emitted in such uncountable numbers, the spawn turns the sea completely milky. The natives know to anticipate the event and complete village communities wade out long before dawn, armed with dishes, buckets and fine nets to scoop up large quantities of the roe-filled palolo tails, considering them a great seasonal delicacy.

Often divers come across long, thin, grey-white tentacles that protrude from holes or under rocks, and searchingly wriggle across the bottom. These are spaghetti worms (*Eupolimnia nebulosa*). The worm lives buried in a soft tube made of sand and mucus, from where it only sends out its tentacles to feed, mostly on bottom detritus. The grooved tentacles can stretch up to 1m (3ft) in length.

Flat worms (phylum *Platyhelminthes*) are among the most spectacular of marine invertebrates, rivalled only by nudibranchs with whom they are often confused – perhaps because they share many of the same brilliant colours and patterns. The most visible distinction is the lack of the gill bouquet and rhinophores found in nudies.

Below: *Flat worms may sometimes be confused with nudibranchs, but they lack the rhinophores and naked gills.*

Wafer-thin and fragile, flat worms crawl over reef surfaces on a self-produced mucus bed, helped by many tiny hairs. But they are also excellent swimmers. When they need to flee or relocate, they do so with rhythmic undulations of the margins of their bodies. Their bright livery, of course, also warns predators of distasteful properties, for they feed on roughly the same kinds of food that nudibranchs do, perhaps just deviating from the specific species.

Delicate and extremely fragile, flat worms are easily damaged and should preferably not be touched, although they can regenerate an entire new animal from just a tiny fragment. This way of multiplication results in a clone, but they normally reproduce sexually.

There are an estimated 3000 flat worm species and the majority reside in the oceans. But flat worms do not preserve well as specimens, and therefore they are by no means adequately studied or photographically recorded.

The acorn worms (phylum *Hemichordata*) are sand-burrowing species that divers never get to see. It is they who produce those puzzling coiled heaps of sandy excrement that we often find on sandy bottoms without being able to pinpoint the creature of origin.

Crustaceans: Packaged in Armour

Crustaceans are truly one of the dominant stake-holders on the reef. Many are microscopically small and form a large part of the nourishing plankton brew that feeds the gargantuan appetite of the reef. They are even part of the diet of giants like gigantic manta rays and whale sharks, which visit the edges of current-swirled reefs to forage on blooming plankton. Across the entire reef, crustaceans feature in the diet of all carnivores at some time or other.

The name crustacean is derived from the Latin word 'crusta', which simply means shell. Like medieval knights in armour, all the crabs, shrimps and lobsters are sheathed in a hard exoskeleton and have jointed limbs. We are all familiar with prawns, crabs and lobsters, but there are about 30,500–35,000 species, of which about 4000 are crabs. Crustaceans live in burrows and caves or under rubble, ledges and rocks. Others spend an entire lifetime in and on sponges and corals, both soft and hard. Some bury themselves by day and emerge only at night, some carry shells, others live in them. Many are permanent swimmers, carried along on the whim of ocean currents and tides.

All crustaceans sport 10 legs, the feature which afforded them the scientific name of decapods. In individual species, the pairs of hinged legs develop and modify according to need; they become pincers or claws, egg-carrying appendages or tweezers, walking legs or swimming paddles.

The hardened external skeleton of calcium carbonate has one drawback: it completely encloses the body and cannot expand. So to grow, crustaceans must shed and replace their old carapace at regular intervals. From the time that the eggs hatch, releasing the microscopic larvae, the animals go through several body-shape changes, adding or subtracting those appendages needed for the particular period of development. Before moulting, the calcium content of the old shell is economically digested or absorbed to resupply the new body shield with strength. The weakened old coat splits and the animal pulls from it, exposing a new soft and wrinkled skin. The discarded 'glove', complete with eyes and limbs, remains on the reef to fool fish and divers alike.

The crustaceans are now extremely vulnerable while their new chitinous soft shell inflates, shapes and hardens. The miraculous strategy of regeneration has replaced previously damaged legs or

Above: *Reef crabs* (Carpilius convexus) *locked in their almost impenetrable carapaces can only mate when they moult.*

Right: *In the dimmer regions of cleaning stations, hingebeak shrimps* (Rhynchocinetes uritai) *sometimes provide cleaning services.*

Below: *The spider crab* Cyclocoeloma tuberculata.

antennae on the new body, and the animal's size is now fixed until the next moult. Throughout this period, crustaceans are extremely cautious of predators and seek deep shelter until they are sufficiently armed for the hostile world again. For some species, this is the only opportunity for mating, as their impenetrable carapaces prevent it otherwise.

Crustaceans are almost always either female or male and, as a rule, copulation is the method of reproduction. Afterwards, the female carries her hundreds of thousands of eggs on a special brood patch underneath her abdomen. She is then said to be 'in berry'.

Extremely important in the food chain, crustaceans are actively hunted. As a result,

they practise an inconspicuous lifestyle and are generally best seen at night. They are well hidden on soft corals, and perhaps this is a good example of how very simple symbiosis can be. On or in sponges, crustaceans benefit from their hosts' filtered food while contributing to home health by removing debris from the circulatory system of the sponge, an act of which the sponge itself is not capable.

Shrimps of different shapes and sizes live in every conceivable crack and opening on the reef. Keep an eye open for the hingebeak shrimp (*Rhynchocinetes uritai*), which can at times be seen in dense colonies on large cleaning stations. Its brilliant red body advertizes unpalatability, which is intensified in a mass, but the shrimp also provides limited cleaning services, and this is the message of its white markings. These shrimps have quirky, jumping movements and make excellent macro-photography subjects.

The sea star crab (*Lissecarcinus polyboides*) lives near the mouth of its sea star host, but also inhabits pleurobranchs. The paired Coleman's shrimps (*Periclimenes*

curled bristles on the carapace, a method that is today successfully copied in the product called Velcro. When such a crab is carefully observed, it becomes evident that it knows beforehand exactly where it is going to place each piece. Once a wisp of decor is hooked on, it gets tugged at several times to ensure that the attachment is firmly in place.

Decorator crabs that use polyps and stinging organisms are skilful precision engineers. They 'grow' the live animals by cutting off polyps as close as possible to their bases and 'planting' them on their bodies with the use of some sort of adhesive. Whether this is from a special gland or present in their saliva is not clear. The fact remains that some special substance

Left: *Night sees soft corals and fans come alive with crustaceans, including decorator crabs* (Naxoides taurus).

Below: *Bits and pieces stolen from the reef soon help this decorator crab* (Camposcia retusa) *to disappear.*

colmani) are only found on the elusive and venomous fire urchin *Asthenosoma*, where they shelter, always aligned head to tail, on an intermediate patch that is devoid of spines, tube feet or pedicellariae.

We hear the many explosive sounds of pistol shrimp claws on every single dive. At night, in artificial light, the thousands of tiny red rubylike eyes can be seen. But it is among soft corals and sponge outcrops that the most fascinating crustaceans can be found. The truly remarkable spider, disguise or decorator crabs in almost all cases favour animal organisms – such as fragments of sponge, hydroids, gorgonians and corals – rather than plants to cover parts of or their entire bodies. Unless they move, these crabs are quite often completely indistinguishable from their surroundings.

Interestingly, decorator crabs that roam widely do not take their decorations from one spot, but use a variety to blend better with the reef. The bits and pieces are collected with remarkable precision and scrupulously trimmed with the mouth. The ends that will be attached to the body are frayed to better hook onto an abundance of

must prevent the rotting of such amputated organisms, for they not only remain alive, but repair themselves and soon continue to grow. Fully complete with a mouth and a crown of tentacles, they are still able to extend or retract the polyps, carrying on a normal feeding life.

The sponge crabs excise living umbrella-like sponge covers and carry these over their bodies with two adapter arms. As the sponge grows, the excess is shaved away. Because the sponges that they choose are mostly distasteful to predators, the crabs are quite safe and very well camouflaged. The beautiful depressed crabs which live on gorgonian corals, although not true decorators, may from time to time cleverly use nipped-off bits from their host fan to 'extend' their bodies for better camouflage.

Hermit crabs are the carnivorous scavengers of the night. They have chosen to avoid the cumbersome process of moulting and have opted for mobile homes – shells which can be exchanged for larger ones as the need arises. The hairy red hermit crabs (*Dardanus* sp.) are the most active and they roam widely, although they have home territories. They not only kill and eat molluscs, echinoderms and other crabs, but also fish. The yellow-and-red striped hermit crab (*Trizopagurus strigatus*) is specially adapted to live only in cone shells; it has a flattened body that allows it to make use of houses that others cannot occupy. All hermit crabs can fold their claws into an impregnable, armoured door once they retract into their homes. The anemone hermit crab cultivates personal live anemones, and does not leave them behind when it moves to a larger house. Instead, it starts tickling the anemones until they release their hold and then transplants them to the new shell!

Above: *Soft-bodied hermit crabs* (Dardanus *sp.*) *derive protection from their shell homes.*

Opposite: *Only their faint markings and swaying movement betray the presence of glassy shrimps* (Periclimenes holthuisi) *on anemones.*

Skeleton shrimps (*Caprella* sp.) are tiny. Mostly found on growths along walls, they hang on with their back appendages, and lean out backwards into the water to catch plankton. But search closely with that magnifying glass, for they mimic both the colour and the stance of their growths.

On the reef floor, mantis shrimps (*Odontodactylus scyllarus*) excavate deep galleries and labyrinths. They have amazingly complex stalked eyes and an aptitude to learn and remember. Strongly resembling the preying mantis, they have specially adapted dactyls, powerful forelimbs which they use for smashing open shells and crabs. They often go for a stroll on the reef, but can also swim by furiously beating their tails and the swimmerets on their abdominal section, paddles which have been modified from feet. The related javelin mantis shrimp has golden eyes and transparent antennary scales; it sports sharp barbed spines on its forelimbs, with which it spears its prey.

There are many more exotic lobsters, shrimps and crabs, so you should never rush past soft corals and sponges. Especially at night, hang quietly, and try to pick up minute movements. With a torch you have a distinct advantage and, keeping in mind where to look, you should soon be able to discover many of the reef's armoured knights.

MANKIND
AND
THE MARINE
ENVIRONMENT

The Marine Intelligence Network

How do we guess at the intelligence of reef creatures? Can we out-manipulate them to get closer? Will we ever be able to make friends with wild animals? Is it intelligence or instinct that enables these creatures to 'know' how to live and protect themselves on the reef?

Undoubtedly, reef creatures, especially the swimmers, are very intelligent, yet we cannot talk about the 'IQ' kind of intelligence. We simply watch the everyday common sense and amazing aptitude for learning that so many reef creatures display. Intelligence is an essential characteristic of all life forms seeking to protect themselves. Do reef animals think? Without question many do! We know that most animals learn from example or by experience. When they don't get what they want, they quickly alter their situation and behaviour in ingenious ways to pursue their goal. To change the rules and their tactics means that they must have the capacity to enjoy, explore, exploit and experience. They simply cannot survive on the cutting edge of chance without intelligence.

Previous pages: *A school of yellowtail snappers* (Ocyurus chrysurus) *surrounds a diver in the Caribbean.*

Below: *Several grey reef sharks* (Charcharhinus amblyrhynchos).

Scientists tell us that an octopus has at least the intelligence of an average cat or dog. We know from experience that reef fish are wary and skittish in areas where spearfishing takes place. If this sport persists, the fish move on to a reef they perceive to be safer. It has been observed that when a diver carries a harmless stick resembling a speargun to a reef that has previously experienced spearfishing, the inhabitants will be noticeably restless and will scatter. On reefs where divers do not molest fish, their initial wariness disappears and the fish become tamer.

In sharks, the ability to connect the sounds of an arriving boat to a given set of circumstances points to some element of 'reasoning'. Divers have noticed that when sharks are habitually attracted to a specific site with food as bait, they will not circle in anticipation if the correct feeding-related actions and sounds are not present. Instead, the sharks remain rather unexcited and aloof. The following experiment was conducted in the Bahamas: one tagged shark was conditioned to press a button, after which a motor sound would start, closely followed by the release of food. It became expert at this. The device was then removed for three months. When the device was reinstalled the scientists were greatly surprised to find that their tagged shark not only remembered the sequence, but was now joined by 10 other sharks who without hesitation knew exactly what to do. How was the information transmitted? Even the scientists cannot tell us yet.

Fish are naturally curious. Drop a foreign object onto the reef, and it will soon be inspected for its potential as a food source or living space. On the many clean-up

WHEN THE LION FEEDS

Confirmation of the existence of a certain level of intelligence in fish was driven home to me during a shallow night dive on a site known to sport five different kinds of lionfish (firefish). Lionfish are nocturnal hunters and have very venomous fin spines which can cause extremely painful wounds. I was warned by a diver, just back from a dusk dive, to watch out for one persistent member of this species, which seemed to be all around him during the dive.

No sooner was I in the water and had switched on my light than I had company. As there was no surge, I wasn't particularly bothered and let the fish be, while I searched for night critters. The lionfish remained steadfastly at the edge of my circle of light. The instant that I saw a small fish, the lionfish pounced. Its gobble-yawn and the absence of the little fish could leave me with only one conclusion. Still, I was not sure, but again and again the lionfish pounced.

Then it dawned on me! This lionfish had observed many night divers, initially probably only out of curiosity. But soon he cunningly learnt that small fish are temporarily stunned by the divers' lights, so becoming easy prey. As soon as darkness set in, the lionfish conveniently waited right under the boat for night divers, whom it now habitually employed in its hunt. It spread its winglike fins like a basket, shading its own eyes from the light and coralling the prey into an effective trap from which it could not escape.

After a while, feeling sorry for the victims because of the lionfish's unfair advantage, I pottered on in total darkness, using my light only when absolutely necessary. Sadly, the diver who warned me had never paid attention to the strange behaviour and missed an amazing and interesting experience.

Above: *Lionfish* (Pterois volitans) *hunting at night.*

expeditions that we join, we see the ever-present junk of man strewn on the bottom: cans, bottles and plastic containers. Often these can no longer be removed, as they have become homes for eels, octopus and crabs who have learnt that they require a lot less maintenance than natural burrows.

We accept that intelligence is a prerequisite for the development of personality as we understand it. Fish and moray eels, as well as octopus, display distinct individual personalities. These marine creatures also actively recognize the hair colour, wetsuit pattern, cameras or dive behaviour of frequent divers. We cannot doubt this, for all too often we have been picked out selectively by favourite 'pet' creatures from underwater groups in which the other members were strangers. While they would remain aloof from the newcomers, they would comfortably stay on or around us.

On a Mauritian reef, we have become friendly with a large moray eel. Eels are normally retiring creatures and often incorrectly perceived to be vicious. For some or other reason, without any specific encouragement or enticement, the eel started following us on the reef. We never made a fuss, instinctively deciding to wait for the eel to call the tune. We did not know whether it had perhaps been fed by others, yet it never engaged in the sniffing and nudging antics that such hand-fed creatures tend to display.

Day after day the eel became less wary and swam ever closer, until it finally dared to twine around our bodies in sinuous ecstasy. Thereafter, as soon as we rolled into the water from our chase boat, our eel friend would swim almost to the surface, inspect our cameras, repeatedly twine itself around us and join us in the descent. Months later, during a separate visit, we were again met and welcomed by 'our' eel, who swam us to its favourite lair to display a shy mate. Although our eel

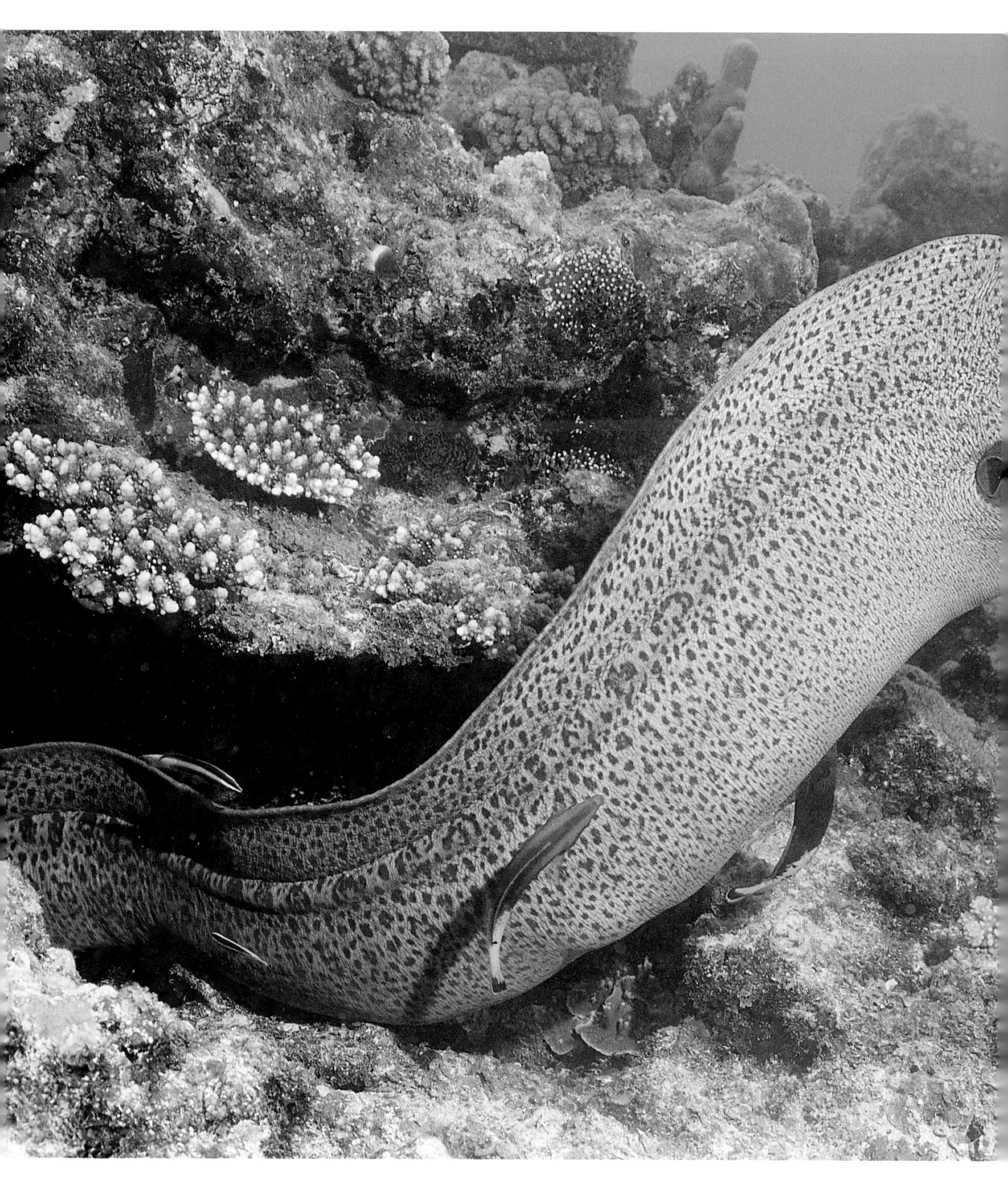

repeatedly joined us, the mate never did. She(?) actually displayed signs of jealousy when she felt the relationship was too intimate, by surging halfway out of her den and nipping nastily at our fins.

Recently I visited the same dive site again to look for our eel, but initially found nothing. After almost an hour in the water, I saw it hidden in a deep crevasse. Nothing could persuade the magnificent animal out of its hide. When I tried to touch it, it scurried away with a very unusual irritated movement, immediately hiding deep in a crevice. I decided to give up any attempt to lure it out and moved on. Soon I switched on the lights of my video camera to film a school of fish. Suddenly there was a poke in my ribs; the eel seemed to have recognized my camera lights. Perhaps the memory that a diver bearing such lights never harmed it set off the spark of recollection. The eel slowly circled me, then came right up for a bit of a cuddle. Unfortunately I had reached the end of my bottom time and had to leave.

Later during the day, I questioned some of the local divers and heard that there was a new divemaster who provided his guests with a macho show. The highlight of his presentation was the vicious prodding of large moray eels with a spiked stick. This evoked not only snapping motions but also surges out of the den, adding a cheap thrill to his excursions. How sad that a diver should be either so ignorant, stupid or downright vicious! As most of the eels were more or less pretamed, he could have provided a much more fulfilling encounter by allowing his guests to pat and interact with it, while he could sell videos and photographs of the memorable occasion. At the very least, his methods drive the eel into hiding; at most, the creature is likely to seek another reef or retaliate one day by biting a guest. The divemaster is a threat to the island's dive industry. All the wonderful dives that once drew plenty of paying guests will become devoid of interesting marine creatures and the profits for everyone in the tourist industry will dwindle away.

Left: *A giant moray eel* (Gymnothorax javanicus).

Danja and I have no magical qualities, but what probably contributes most to our outstanding experiences is our level of comfort in the sea. Scuba skills have become second nature and we both feel as comfortable underwater as we do on land. We have honed our equipment down to the bare necessities. We have worked out which clips attach equipment effectively and have become as streamlined as seals. Through diving extensively, we have gained much knowledge about creatures and their lifestyles, and can often anticipate many interesting events on the reef.

It has become second nature to consciously consider every approach, evaluating all the factors in a given situation, before making the decision to approach an animal for study or photography. Naturally an instant reaction is required for some of the unexpected events one encounters on reefs. It is the patient practice accumulated during many previous encounters that will provide the almost instinctive and correct reaction in unforeseen moments, when things happen with lightning speed.

Below: *A diver explores a coral overhang.*

Safe Diving

In diving, as in any other sport, there is the potential for injuries to occur. However, diving risks can be calculated and prepared for; they are overcome by sufficient skills, knowledge and equipment. Each diver must be able to state before every dive that: 'This dive will be safe for ME'. But, in order to do so, there are a few points divers should always bear in mind:

- Re-evaluate your skills and attitude at regular intervals. Never stop questioning the validity of the lessons you learnt in your course and keep up to date with new techniques and equipment.

- Know and practise emergency self-help skills. Although one should be, in truth one is not always 'within touching distance' of one's buddy. For this reason it is a good idea to carry a completely separate air supply with its own independent regulator.

- Learn to be qualified enough to solo dive. This certainly does not mean that you **should** solo dive, but the only truly responsible and safe diver is one who learns **as if** he will dive alone. Such divers check their equipment more diligently, dive more conservatively and get into trouble less often. They are also capable of handling emergencies and do not take chances.

- Finally, never get too cocky for your own good. NOBODY is above the rules of diving safety. Follow the rules all the time and know your limits. If you feel physically or mentally uncomfortable on a dive, GET OUT – the reef will still be there tomorrow!

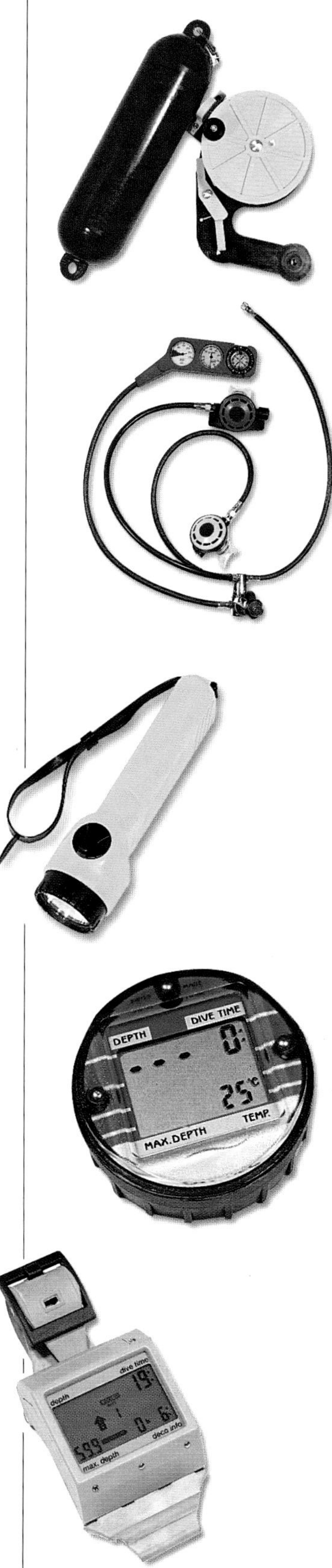

How to Forge Marine Relationships

It is a fascinating reality that we can get much closer to wild animals underwater than we can ever hope to do on land. While any interaction with land creatures results in an instant choice of flight or fight, most marine creatures are gracious enough to remain almost within touching distance. And yet, we cannot fail to notice how some divers always get closer to fish and reef creatures than others. What mystical qualities do they possess?

The truth is, these divers simply have 'ocean attitude'. Far from announcing themselves as intruding visitors, they are keen observers who take every possible measure to become one with the reef environment. They are calm and controlled. They build animal trust. Their consistent good luck is merely a product of being well prepared for opportunity. Danja calls it 'being one with the sea'.

Unconditional acceptance by a wild animal is a rare experience. Some of us will perhaps have one chance to experience it, some never will. But any wild animal interaction must always be based on two criteria: it must be positive for both species and it must be mutually enjoyable. Always carry fairness and respect with you into the underwater realm.

If you have ever been underwater while other divers take the giant stride, you will have noticed how fish wince and duck for cover or flee at the moment of impact. While divers learn a variety of entries into the water, they never again think about the validity or suitability of these methods. Many entry techniques originated in the navy or were developed for boats that were never intended for diving purposes.

Yet, these styles perpetuate unchallenged, although every other aspect of diving has been refined technically and environmentally. Dive-boat operators usually anchor at the edge of reefs, where most reef animals are and where much marine interaction takes place. Modern dive-dedicated boats have comfortable surface-level dive platforms that make a quiet slip into the water possible. Even the well-known backward roll off small boats can be staggered to reduce impact and with a slight body twist may be quieter and more controlled.

The development of consumer rebreather apparatus has been eagerly awaited as divers assumed that the consequent lack of

In Their World on their Terms

(Motto of the Wild Dolphin Project)

Recently, Danja was swimming with dolphins in open ocean. Several other people were also in the water. On snorkel, most hung in the water with the typical surface-oriented 'dead man's float'. But for the best part of 40 minutes, Danja dived down repeatedly, cajoled, twisted, looped-the-loop, and played with several dolphins in one of the most haunting interaction ballets I have ever seen. Analyzing the event on film much later, I noticed that she never once reached out to any of the animals. Instead, she imitated their movements, held still for their sonar inspections, giggled back when they clicked their messages and looped in giant circles to attract them to play. Finally she experienced that magical moment when the dolphins voluntarily and continuously swam belly to belly with her, happy to be hugged and held, while she equally happily released them at the slightest indication that they had had enough. None of the other snorkellers attempted to emulate these actions. To them, Danja simply had some 'mystical magic' that caused the dolphins to be besotted with her. We can all learn to have wonderfully satisfying experiences like these, but to learn we have to give considerable thought to all our desires and actions.

Above: *An encounter with the nautilus* (Nautilus pompilius), *the primordial ancestor of molluscs, is truly memorable. The nautilus lives at great depths, but ventures into the shallows at night to feed.*

bubble emissions would make them less noticeable. But professional photographers who use the apparatus extensively in order to lengthen bottom time, mostly in a controlled fixed spot, now report that there is no marked difference in being accepted on the reef. Moreover, its bulk and the technical complication of mixed gas at different depths makes it more of a hindrance than a help, especially when a diver wishes to move around at different levels on a reef.

Snorkellers need to acquire quiet swimming skills and must learn to propel themselves with their fins only. Very often a slight tip of the body towards one side will help to keep fins underwater for effective forward propulsion and control, instead of noisily slapping and splashing on the surface. Snorkellers should also opt for a wetsuit or dive skin – valuable protection against stingers and coral scrapes, as well as the sun and the cold that inevitably sneak up when one floats quietly.

The Importance of Developing Trust

A vital element in achieving close encounters is trust. Marine animals have no way of understanding that you mean them no harm. Only by carefully watching the signals you send can they evaluate what threat you represent. Trust and rapport-building is perhaps time-consuming, but it is absolutely essential for the diver who wishes to have fulfilling experiences with marine creatures.

If it's spectacular marine photography you're interested in, this calls for very close proximity. Subject-to-camera distance is crucial and must be drastically cut underwater, due to the filtering effect on colour and light and the many suspended particles. Thus, photographers must learn or

Above: *When initiated by the dolphin, close encounters can be magical.*

Right: *The ornamental coral cod* (Cephalopholis miniata), *or coral hind, is an attractive but shy species that lives in caves and hides under ledges.*

develop techniques and tricks to succeed – techniques that can also greatly benefit the nonphotographer too.

Naturally, different techniques apply to different species. Sessile and slow-moving animals present few problems, although clams and the beautiful featherduster and bottlebrush worms retract instantly if they sense a change in light or water pressure. Adapt your style of approach accordingly and don't make waves. If they still retract, wait. With enough patience, you may see them venture out again. Videographers, steady your cameras and keep focused – their emergence makes for ethereal footage.

Fish are probably the most difficult of all reef creatures to approach closely. For photography you need them out in the open, visible and preferably still. It is important, therefore, to be aware of 'fish trails', invisible roads or paths along which there is regular traffic. Find some sort of coral structure along these trails that will provide cover. If you are a photographer, choose the structure for its attractiveness as a background. Now begin to observe the habits of the passing fish, but try to appear disinterested and seemingly avoid eye contact.

Initially these agile and elegant movers will be suspicious and your intrusion will be warily watched. Remain calm and still, preferably at a low level. Establish a calm breathing rhythm, with slow and even exhalations to avoid overnoisy bubble spurts. Do not poke or point at an animal that is stationary, as it might take quite a while to make up its mind to trust you. Fish may flee or retreat at first, but they are incurably curious and will soon return for a second look. The longer you can be patient, the luckier you will get. As soon as the fish conclude that you are probably not a threat, they will accept your presence and resume normal reef behaviour. This method facilitates your accurate anticipation of recurrent appearances – it lets photographers prefocus and capture faces instead of fleeing tails.

When you approach any reef animal, remember that they have extremely acute senses! Every advance should be slow and always performed from a perspective that will allow the fish or creature to see you. Watch the eyes and fins of the animal constantly. They clearly indicate when you overstep the invisible line. The slightest sign of intense or overeager pursuit will be thwarted repeatedly. Rather retreat slightly and hang around calmly. Remember that most reef fish are quite territorial – many are at least home-ranging – and will return as soon as they feel comfortable with the level of your respect. Pretending to look at other things usually works! Fish seem to have huge egos and they may insist on being in the limelight.

Watch for signs of restlessness as you edge closer bit by bit. Stop and wait if fish lift their dorsal fins. While there is a definite limit to the proximity you can enjoy with any wild creature, I have found that it depends much on pressures of predation. The unwritten borders may vary from day to day or even hour to hour, but crossing that invisible line will instantly break any trust that may have been established. The more time you invest, the sooner normal behaviour will resume and the more accurate will be your evaluation of the limit of closeness. I have waited many times with spectacular species that were anticipating being cleaned, or with fish and octopus I suspected were courting, and they always proceeded in front of my eyes as soon as they noticed that I stayed still.

If an animal flees in alarmed panic, you may just as well find something else to occupy you. Pursuit will only cancel any chances of success. Frequently an animal that was frightened once but not pursued, will calm down and take its cue from other more placid and curious inhabitants. You may then still be able to spend time with it later during the dive. This building up of a history of trust is also the reason why repeated diving on the same reef is often coupled with progressive success.

Below: *Divers must stop and wait patiently for reef life to resume its normal routine.*

Always match your placement to the level of the fish or reef animal you wish to observe; if possible, find a level slightly lower. Not only is it the better vantage point from which to photograph, it is also physically more comfortable and your size appears less threatening. When you loom above marine animals, you are perceived as a possible predator, and everything will flee or dive for cover.

Marine animals are experts at reading body language. Tense bodies and hand and fin motions may all be interpreted as signals that indicate safety and control or panic and danger. Interaction should not be initiated without some clear indication that it will be accepted or welcomed. Any relationship is pointless if the signal does not come from the marine creature.

There is, literally, a rather silly-sounding technique that successfully taps the natural curiosity of marine creatures. If you hover or cruise horizontally, like most underwater animals do, the wet set inevitably move in to inspect you. When this happens, it pays to become more interesting. By crooning or humming into your regulator, you will often draw a curious audience that tries to make sense of this impromptu performance. Humming works particularly well for sharks, notorious for their disinterest in divers, and they will frequently stay around for a while to puzzle at the origin of the sound.

Below: *Good buoyancy control is essential around delicate reef structures.*

Turtle Tactics

The strategy of ignoring creatures is quite well known among diving photographers. Apparently, sea turtles are especially susceptible to such subtle insult. I have been told that the idea is to immediately look away when you see a turtle on the reef. Somewhat later, glance at it fleetingly, but continue your surveillance, noticeably past the creature. Studiously keep on ignoring the turtle, affording it only short, totally detached glances. The turtle soon displays annoyance and attempts to attract your attention. Continue feigning disinterest. The turtle finds this disregard so unbearable that it advances almost into your face, insisting on your attention.

On the few occasions that I could test the theory, it always succeeded. How regularly this will happen, I cannot say. Certainly 'ignoring' other creatures on the reef has often worked for us and we employ the ruse on almost every dive, particularly if a subject is uncooperative.

Patience Produces Results

Photographers, often confronted with challenging technical problems and loaded with bulky camera equipment, need many more skills than unencumbered divers but do not always possess them. Most of the time photographers are criticized for their handling of fragile creatures to improve photographic composition. This unpleasant practice is often caused by impatience and is often completely unnecessary.

It is easy to damage fragile creatures such as nudibranchs and flat worms, so please take special care and be an ecofriendly photographer. There are many gentle ways in which temporary relocation of a marine animal can be accomplished, but always consider relocating yourself first. A well-aimed beam of light helps very effectively to 'direct' nudies and flat worms into more aesthetic positions, without you having to touch them. As a rule, rather push very gently than pull, especially with limbed creatures, who will often more willingly sacrifice a leg or antenna than be 'captured'. If creatures are moved – and remember you will be criticized – they must be replaced in the exact spot they were found. Ecological niches are extremely finely divided on the reef and there is fierce competition for and protection of space. So each hole may be an occupied territory or present other dangers.

Photographers in quest of very close-up fish-face photography should know that there is always, in divers' lingo, a 'dumb fish' on the reef. These seem to be rather slow-thinking and less prone to panic, often posing beautifully while evaluating circumstances and planning the path of flight. The stunning results feature on the front pages of many dive magazines.

Top left: *Patience is the key to any close encounter.*

Top right: *This coral trout* (Cephalopholis miniata) *poses beautifully.*

The large triggerfish is one of the species you will often be warned not to approach. The reason is the zealous and extremely aggressive guarding of the egg nest by the male. But it is easy to evaluate whether brooding is actually taking place. Short forceful rallies of circling swims interspersed with repeated puffs at the same location, coupled with fierce chasing of other fish in the vicinity, spells danger and indicates that the triggerfish is guarding eggs. It is then extremely dangerous and will attack intruding divers. The safety margin extends well beyond the distance the fish covers when chasing intruders away. Triggerfish are capable of inflicting really nasty bites and they may cause enough panic in a diver to result in other problems. Outside the breeding times, triggerfish can be safely and quite closely approached. They also huff and puff at different surfaces when they feed, but this should not be confused with brooding behaviour. At this time, the fish does not circle aggressively or remain stationary for long, but moves on constantly.

Lionfish are one of the superbly attractive venomous species suitable both for striking close-up and wide-angle photography. As they are nocturnal hunters, they tend to remain low on the reef by day, hanging almost motionless while looking quietly majestic and haughty. Yet, if photographed from above, their splendour cannot be separated from the confusing reef background. They rise higher in the water column during the late afternoon, which is the best time to photograph them. However, they have one trait that always undermines their determination to stay aloof: they are extremely inquisitive! There is simply no better way to attract them than to appear studiously involved in something else. Pretend to be watching or photographing another subject close-by, exaggerating the intensity of your interest. Ignore the lionfish completely. It unfailingly rises and approaches you from above to investigate, perfectly positioned for a beautiful shot.

These fish carry their venom for self-defence, not attack. They may be approached very closely, provided that you are in control of your body. In a surge you need to be experienced enough to swing backwards and forwards with the rhythm of the sea, duplicating the sway of the lionfish. If all your movements are gentle, there is less danger of impact and an accidental brush with the fins will not necessarily cause penetration or the release of venom. Novices should practise on harmless species!

Photographers must anticipate that special 'peak of the shot'. Apparently disinterested lionfish often give huge mawing yawns which, when shot head-on, result in fearsomely impressive winning shots. The onset of 'chewing' mouth movements heralds this behaviour. Videographers should then record continuously, while still-photographers must be preset, prefocused and with finger permanently on the release button to catch the exact moment of the yawn.

Opposite top: *The giant triggerfish* (Balistoides viridescens) *can be closely and safely approached when it is not breeding.*

Opposite bottom: *A clown triggerfish* (Balistoides conspicillum).

Below: *'Bored with the photo session?'*

Photographers are always on an unceasing quest for the 'perfect' clownfish shot. Clownfish are initially very agitated when they are approached. The whole community will yo-yo up and down or in and out of their anemone home, never staying still long enough for you to compose a picture of any significance. Residents of sand-burrowing anemone species seem to get particularly excited, perhaps because they fear the loss of protection should their anemone be disturbed and retract.

Be patient in your approach. Quiet hovering helps to reassure them quickly. If clownfish are particularly frantic, they may have a brood of eggs under the flap of the anemone, which they will protect fiercely. These damselfish are incredibly courageous and not above lightning-fast attacks in such cases. They have sharp teeth that can bite quite a chunk out of a diver, and they unfailingly choose unprotected body parts like faces and wrists. Some clownfish may even flee their anemones temporarily, but they never venture far. If they feel confident that you are no threat, they soon settle into more predictable movements.

The larger the clownfish community is, the greater the challenge to get all the fish in focus. This is not easy, but certainly possible. The fish are habitually fast and quirky, and separating in different directions is a rather old and proven distraction strategy. Do not follow their movements with your camera. Rather compose a well-framed shot, lock focus and, finger on the button, allow the fish to move in and out of the picture. Soon the desired, more composed behaviour will begin to happen. Patience is always rewarded. If you fail, you simply have not followed instructions!

When creatures like octopus are safely in their burrows, it is difficult and perhaps even unreasonable to lure them out. It is especially important to check that no hungry eels are around. But octopus are very curious and it is easy to hold hands with them! Very often, all that is required is a finger held completely still at the burrow entrance. When an octopus moves on the reef, freeze and keep your eye on it all the time for unusual behaviour. The octopus will watch you equally closely and with much greater patience. If it is not alarmed, the octopus usually resumes whatever it was doing before. Your success is related only to the time you invest in motionless anticipation – hypothermia is a real possibility, but it is wonderfully offset by the warm glow of victory afterwards!

Below: *Holding hands with an octopus can be a thrilling experience.*

INTERACTING WITH THE 'BIG ONES'

One golden rule dictates all close encounters with the big creatures. However irresistible it may seem, never ever swim towards them or stretch out in an attempt to touch. While the 'big ones' – dolphins, manta rays, sharks, whale sharks, big groupers and seals – will leisurely tolerate sharing their space with you and might even venture close for a thorough inspection, they always move off instantly if contact is attempted. Most dolphins and whales dislike diver bubbles, perhaps because bubbles for them are a sign of aggression. This makes snorkel techniques a prerequisite. While swimming with whale sharks is pure magic, it is also exhausting. Such encounters almost always take place on snorkel only, and quite strong swimming skills are necessary to keep pace with these giants. Manta rays are very shy and while one could in isolated cases attempt to swim with them, it is probably always better to wait for their approach. When mantas flee, they do not return. The best way of seeing them is on the current-swirled reefs they regularly visit in order to feed or be cleaned. For most of us this will entail travelling to 'guaranteed' locations such as Yap in Micronesia.

Many potentially magic encounters are sadly and prematurely ended by clumsy attempts to interact. These animals are acutely aware of your presence and their curiosity is as strong as your own, but each encounter will be on their terms only. Even on underwater vehicles, the ease and speed of the big creatures in their natural habitat is without doubt superior to ours. Now is the time to remember the humming technique! It is a proven fact that encounters last much longer when divers practise restraint. Even when the creatures are very close, avoid reaching out. Their approach merely means an inspection, not an invitation to interact.

You may be able to touch tamed creatures – such as eels or stingrays and perhaps some dolphins – on reefs where creatures have been fed and conditioned to interact. Please do so with permission

Above: *Marine life posture clearly indicates when interaction is permitted. These are the unforgettable moments of sea friendships.*

only and follow the divemaster's instructions to the letter. The rules were made to ensure protection for both divers and creatures, and to allow for optimum interaction. Please do not wear gloves, they are unneccesary in the tropics and extremely abrasive to the creatures' silky, mucus-covered skins. Besides which, gloves rob you of a delicious and, for once, permitted tactile pleasure. Consider the consequences of your actions carefully and be courteous to both the creatures and your fellow divers. Headstrong satisfaction of your own impulses will spoil other divers' chances for observation, photography or interaction and make you a diving pariah. Remember that your divemaster is authorized to ban you from all his and his colleagues' future excursions.

For the best view and photographic opportunities at popular shark-feeding dives, everyone will want to be placed near the bait or as close as is safely possible. In spite of many express rules and undertakings during predive briefings, these experiences very often turn into chaos. The larger the group attending a shark feed, the more unpleasant the dive becomes. Polite divers seem to see nothing but the bubbles of others, and photographers often have regard for no-one but themselves. Discuss photographic needs with both the divemaster and fellow divers beforehand. The best time for good shots is immediately before feeding or after the frenzy has passed, when sharks may circle elegantly for quite a while.

If it is at all possible, serious photographers should purchase exclusive or at least diver-limited shark-feeding outings. Although this may be expensive, the resulting experiences and photographs are superior to those on any shared dive. Do read up on shark behaviour beforehand. We agree that

sharks are magnificent creatures, but we should be aware at all times that we share their world, not they ours. In unnatural feeding situations sharks are unpredictable and it is essential to be cognizant of their danger-posture signals. Mistakes can be made by both feeder and animal, but as almost all shark-obsessed divers will admit, they alone were always the cause of unpleasant experiences. We are certainly not natural food to sharks. If our actions and movements are correctly executed and we refrain from sending prey or challenge signals, we are just another large sea animal to them.

On some reefs in the Maldives, grey reef sharks are as much a part of the scenery as damselfish. When divers descend to about 30m (100ft), where they habitually cruise, the sharks remain at that depth. But when divers stay at a depth of 20m (60ft) or even 15m (50ft) for more economical air consumption, these inquisitive sharks move higher up in the water column.

Predatory strikes on the reef usually occur during the twilight hours at dusk and dawn. At the end of a late afternoon dive on one such reef, I whiled away my safety stop in barely 5m (16ft) of water. Here, while watching courting octopus, I repeatedly saw a grey reef shark swooping in and catching its prey in a single smooth, hydro-dynamic glide. The execution was so superbly elegant that not one fish twitched in alarm. Only after repeated strikes did some species seem to become aware of the shark and even then their flight was not particularly frantic. I could not help comparing the elegance of natural predation with the dramatic, unnatural feeding frenzies so often seen on popular reefs or portrayed in films. Consequently human-induced shark-feeding experiences never held the same thrill for me again.

Above: *A well-controlled shark feed at Walker's Cay in the Bahamas presents plenty of photographic opportunities.*

SEEING THROUGH CAMOUFLAGE

Finding almost invisible shrimps and crabs in soft coral, on sea fans and whips or in commensal hosts is just a case of knowing where and how to look. This is where a knowledge of camouflage strategies really pays off. By studying the books of the world's famous underwater photographers, you can become aware of the existence of these creatures and learn to recognize their habitats.

I have often found that an awareness of a creature's existence alerted me to its presence much quicker than all other information put together. You may frequently recognize the particular home first and only then find the creature. Field and identification guides for specific areas may not contain spectacular photographs but they give vital information on which indigenous species to expect, at what general depth and range and of course, on habitat and food. I jot down such information in a small notebook which I refer to when I dive, especially on new and unfamiliar reefs.

A word of caution, though. I have repeatedly found that depth indications can be very misleading. Conditions on reefs are very diverse and what is normal on one reef may be exceptional on another. Sometimes creature-harbouring organisms cannot find their usual niche and make compromises or adapt, providing all other conditions are favourable. For years, I searched for the longnose hawkfish which lives in gorgonian fans. In the Red Sea they are consistently found at a depth of 20–30m (65–100ft). On other reefs I could never find a single one. My first encounter with this species elsewhere was at 5m (16ft) in a scraggly black coral on a Fijian reef. Once the depth theory and thus the usual habitat was invalidated, I searched gorgonian fans and black corals on much shallower reefs and consequently found the fish regularly, mostly at around 10–15m (33–50ft). Papua New Guinea's reefs are known for unbelievably rich life, unexpected habitats and miraculously exotic creatures. On these reefs 'theories' are turned upside down during virtually every dive. Here, barely 1m (3ft) away from the coastline, I found a giant gorgonian fan and its tartanned resident hawkfish pair at the improbable, unbelievable depth of just 3m (10ft)! Moral of the story: if the habitat is present in shallow waters, there may very well be a few exceptions to the rule!

Below: *Two fingertips on a dead patch of coral stabilize a diver effectively; it is also environmentally friendly.*

Above: *Currents may transform a reef scene beyond recognition within minutes.*

Taking on the Currents

Many divers intensely dislike currents, others are used to tranquil waters and have never learnt the techniques necessary for current diving. But of course the action is always where the current hits the reef and the presence and true splendour of several creatures and organisms is revealed only there. If you feel unsafe or scared in these conditions, find a suitable teacher to initiate you into the tricks of moving in real currents. This kind of diving equals the thrill of sky-diving, without the risk of a neck-breaking fall. However, suitable backup arrangements must be in place and double-checked. On open-water reefs, arrangements must be made with chase boats for pick-ups, and the wearing of those nifty inflatable safety 'sausages' and whistles becomes obligatory. Enter the water only if you possess the skills for problem solving, the necessary safety gadgets and reliable backup.

If, instead of fighting against the stream, you have the opportunity to be dropped off at the edge of the reef that receives the full blast, the first part of your dive is best spent staying quietly tucked in behind an outcrop. Keep resistance to the current to a minimum by minimizing the bulk of your body. We call this 'making like a nudibranch, not like a gorgonian fan'.

Reef creatures seem to arrange a temporary truce at this time, and in many species, behaviour changes markedly. For coral reefs, currents equal bouillabaisse. In the rich planktonic soup, soft corals plump up and every polyp emerges. The fish place themselves head-on into the current with a sense of urgency. It's dinner time! Between snacks, fish hold still for cleaners. Sharks, tuna, barracuda and trevally tend to swoop in or cruise by to take stock. In the frenzy of gorging themselves before the slack, the fish are less tense, enabling a much closer approach. In fact, for watching fish action on any reef, always select that part where the current, however small, impacts with the reef.

If your boat is anchored down-current or if pick-up skiffs have been arranged, end your dive with that most blissful of all experiences – drifting. Use the current as your personal magic flying carpet and – instead of observing the individual – see the entire, interrelated, living and functioning reef. Never forget to look consciously at the whole in your quest for those special creatures.

Photographic Tips

However often photographic advice is given, one always forgets some of the most crucial points. Today's advanced cameras have taken care of many cumbersome problems like bracketing and exposure metering. 'Gifts from the sea' happen suddenly and always when least expected – this is not the time for fiddling with camera controls. It is sensible to have cameras preset for general conditions and ready to fire, so work out a practical system for yourself. Photographic creativity and style is an individual thing but technique and preparedness are essential skills if you wish to attain a predictable measure of success.

For spectacular results, it is better to concentrate on getting one scene perfectly, rather than chasing down a series. Once you find suitable subject material, vary your angles and distances, move in close. Do not forget to show habitat, interesting behaviour and even diver-reaction. Hold your camera still! Leave nose and head room and let your subjects swim into and out of the frame.

For video, leave ample preroll and tail ends. Fill the frame. Zoom with your fins, not your camera. Shoot at an upward angle. Separate subjects from the reef. Let the subject, rather than the camera, move. Haphazard bits and pieces, even if shot on a rich and enticing reef, never tell the story satisfactorily and make editing a tedious, if not impossible, task. In edit rooms such footage is disrespectfully called 'just tape'. Unplanned photography results in boring movie catalogues of creatures seen, rather than riveting adventures. Watch and copy the photographic style of the professional greats who concentrate on just four or five perfectly shot interlinking scenes to create a glorious story that keeps the audience entranced.

Stories that depict marine life must be self-contained and informative; the most successful ones depict behaviour. Emerge from each dive with only one good story rather than with several incoherent scenes. It is amazing how quickly one builds up a decent library from which to tap later if one concentrates on all aspects of a single

▲ *One should always make sure that the subject matter fills the frame.*

◀ *The photographer must remember to move in and isolate the subject.*

encounter. I write short wish lists of footage I need or covet and add subjects that will complete or 'pad' previous shots. I also note down or sketch those shots of other photographers that I find brilliant and from whom I can learn. I read these lists religiously before and between all my dives.

Every time I edit video footage, I religiously add my wishful 'if onlys' next to my 'gee whiz, that's nice' remarks in my notebook. This little book always lives on my editing desk when not travelling with me. I note styles or angles that were super successful and I repeat them underwater. But I also jot down reminders and admonitions, like changing my position to separate my subject from the reef, showing it against blue water, filling the frame, or shooting, then moving and shooting again often enough to have a choice of 'best' compositions. The latter is especially valid for spectacular fixed organisms like soft corals, fans, whips, crinoids, squirts, sponges and structures which helpfully stay fixed in place. These are the scenes that often add the 'oomph' to any final result. Of course, like all other experienced photographers, I also make mistakes every so often. I forget not to point my camera down at the reef or, even worse, to switch it off when it is pointed nowhere at all!

All divers know that some dives are simply 'jinxed'. Exciting things may happen all at the same time or in such quick succession that all previous thoughts of discipline are erased. Enjoy! Other dives may be riddled with obstacles like rainy weather, currents, inadequate visibility or plankton 'blooms'. Rethink your approach. Bad visibility means lots of food for reef creatures. These are the very best times for extremely close-up shots of small subjects, and it is prime time for videographers. Water always appears cleaner on tape than in reality. In these conditions you have the advantage as there is less pressure to share subjects with the stills photographers. There is really and truly no such thing as a boring dive, there are just blunted and unseeing divers.

Perspective is everything. Here, Danja aims her camera into a cave filled with fish in order to enhance the composition of her picture. ▲

It is important to choose a colourful subject and to make use of natural lighting by shooting upwards to include a sunburst if possible. ▶

Preserving the Paradise

Moral responsibility on the reef is not only about conservation. It is a diving philosophy that has everything to do with the dignity and respect with which we enter this very last, almost untouched wilderness on earth.

In fact, divers themselves are responsible for less than one per cent of all reef damage. Most ecological problems encountered in the oceans originate irrefutably from unsound environmental behaviour on land. By now we must surely all have realized that minor rearranging or repair to the environment is no longer an option.

Nature, herself, at times also becomes a destroyer. Any diver that has seen the devastation wrought by cyclones knows what real underwater ruin means. Corals are smashed and crushed to rubble, debris and sand stiflingly settle on everything – and yet the reef survives and in time regenerates, often more beautifully than before.

Most divers are besotted with, and fiercely protective of, the pristine purity of coral reefs. Selfishly, they are glad that a relatively small percentage of the population regularly utilizes reefs for recreation. Most are ecologically aware and involved in ongoing conservation projects, cleaning up after many thoughtless exploiters of the oceans, removing plastics, balls of discarded fishing line and abandoned monofilament nets. Our contribution to oceanic protection by reporting on deterioration or fighting for marine parks, moorings and better marine husbandry is simply priceless.

So, how do we feel about touching things underwater? Our tactile sense develops very early and plays an important role in our exploration and perception of the world. Yet touching is a hotly debated issue and remains controversial.

Many people feel that the blanket rule of 'Do Not Touch' should be the only rule that applies. However, this rule is negative and patronizing, and typical of orthodox ecological thought. When seen in the light of cyclones, storms, crunching parrotfish and the devouring crown-of-thorn stars, it will simply never convince divers. Neither will it be practised consistently. Research has shown that, however much admonished, divers on average touch the reef at least seven times during an hour-long dive. I would therefore prefer that divers be taught when and where touching will do no harm.

Touching in itself is also not really the problem. It is the associated desire to control what we touch that is harmful. While most divers are not malicious, many certainly lack self-restraint or the appropriate manners for interacting with wildlife. Thus, re-education becomes the backbone for changing

Previous pages: *Underwater tranquillity is the diver's spiritual food.*

Above: *Diving with my soulmate along a reef wall.*

Opposite top: *Body control helps a diver to hover calmly near otherwise skittish reef inhabitants.*

Opposite bottom: *We hold the future of our reefs in our hands.*

incorrect behaviour. Touching remains a privilege that needs responsibility, decency and reef-knowledge, together with the learning of appropriate skills. The diving fraternity, including diving scientists and ecologists, have thus formulated guidelines that are based on scientific fact as much as on environmental caring and enjoyment.

Developing Diving Skills

Neutral buoyancy heightens diver confidence, control and water comfort. It also drastically lowers air consumption. Neutral buoyancy gets us much closer to reef life and almost completely eliminates the necessity to make contact with the bottom. It is the single most desirable dive skill, easily practised and taught. If every seasoned diver would pass on only this one skill, popular reefs would look decidedly different. The resultant weightlessness provides effortless control and balance, even in surges or swells, enabling divers to lift gently up and over even the most delicate corals with one simple deep breath.

There is one more remarkable aid for momentary stability – your fingertips. The gentle pressure of two fingertips on a dead patch is amazingly effective for steadying a diver, especially when you are photographing. Using them as miniature 'legs', you can gently walk on fingertips over fine silty or sandy reef bottoms, without destroying burrows or stirring up clouds.

It takes only practice to evaluate the additional length that fins add to your body. While the body should be kept aimed away from fragile reef structures, knees should be gently bent and fins should point slightly upwards. Soft flutter kicks propel a diver forward gently, yet as effectively as stronger strokes, and do so without damage to fragile corals or areas like sandy and silty bottom surfaces and meadows. The epitome of diver skill – motionless hovering – seems to be the most difficult technique to learn. The trick is a combination of three things: correct weighting, correct inflation of the buoyancy control jacket and, most importantly, the cessation of all finning. To simply stop finning is one of the most difficult tricks for novices to learn and yet once it is understood, all other neutral buoyancy skills almost automatically fall into place.

With the above skills, we eliminate two of the three ways in which we touch reefs, namely with knees and fins. This leaves our hands. Gloves are not necessary on tropical reefs. In fact they very often lend a false sense of security and can become a real problem amidst an array of venomous and exquisitely barbed creatures. Spines caught in the weave of gloves cannot easily be removed, and may even aggravate injuries in the process of removing the gloves. Bare hands are the best deterrent against a temptation to touch what is better left undisturbed. At the same time, bare hands also transmit the incredibly silky feel of reef animals like eels and stingrays.

THE ETHICS OF TOUCHING AND FEEDING

It is only acceptable to touch reef animals and plants gently and noninvasively. To be a harmonious part of the oceans, we should touch only to experience, not to possess or show off. While I like to remind myself that the disturbance of one grain of sand may destroy a tiny universe, and that destroying a tiny universe may affect the whole, the reef is not a seamless, living carpet. There are always places where one can stop off without crushing any living thing.

Think and observe carefully before you connect with the reef. An awareness of organisms, their habitats and the roles they play must become a conscious part of every diving experience. Without patient observation, we simply cannot know whether an animal is perhaps resting, breeding or trying to hide. However innocently intended, a heedless act may disturb or, worse, destroy the very object of our interest or the system of which it is a part.

Always avoid damage. It is both common sense and simple to refrain from using corals as handholds or tuck-in spots. Corals are living animals, not reef furniture. They are largely defenceless against abuse from inelegant divers. When touched, corals emit slime. In part this slime is a disinfectant, but it also contains the guts of damaged polyps which have been cut against their own tiny sharp-edged cups. A damaged spot can rapidly become infected, causing coral death and the loss of a home for many creatures.

Below: *Consideration and restraint are the magical combination for getting close to underwater creatures.*

Keep indiscriminate handling of marine creatures to a minimum and be gentle. Resist the temptation to move fragile, slow-moving creatures. Always remember that an animal chooses its location with great care for optimum safety, adequate food and suitable reproductive conditions.

Remind yourself that interaction is only fulfilling when it is welcomed or initiated by the animal. Ask yourself how you would feel if a complete stranger jumped on your back to hitch a ride. Translate this to turtles, mantas and whale sharks. Consider the distress of pufferfish scared into 'blowing up' for fun. An octopus, innocently removed from its house, may in an instant become food for an alert and hungry eel. We have seen this happen in the blink of an eye. Of all creatures, dolphins perhaps teach the best lesson. Although they almost never allow actual contact, there is an exchange of communication that is unbelievably profound – these are soul friendships.

We are all tempted at times to feed marine animals. But how ethical is feeding? Trying to 'tame' a wild animal with food might result in tragedy. Not only may you induce frenzied and unnatural behaviour, but fishermen, spearfishermen and aquarium collectors very often share reefs with you. A tamed and people-trusting animal is conditioned to suppress natural fear, a defence without which its life is in real danger.

Alien foods are frequently used to lure creatures closer. I have seen cheese, hot dogs, bananas, peas, and bread used extensively. If these foods were acceptable, they would certainly be naturally available underwater! Our choice of cost-effective and easily obtained food may in the long term bring illness to the reefs. At the world-famous Red Sea marine park, Ras Mohammad, the statuesque Napoleon wrasse may now no longer be fed. Thousands of boiled eggs found their way into the creatures' stomachs and ultimately their livers – it is unthinking divers who are individually and collectively responsible for the many deaths among this wrasse population.

I accept that a fishy titbit might afford a good photographic opportunity, but even then carefully consider every decision to feed. If food is used to lure animals out of dens, they might unwittingly be exposed for long enough to be caught by a predator or to lose their brood. Feeding-induced jealousies and mistakes are often the cause of diver injury. As photographers, we have found that patience is always rewarded with success and we now largely refrain from introducing any food.

We also abstain as much as possible from moving creatures to colour-coordinated backgrounds. This is an old-fashioned, 'arty' practice, today much frowned upon. If we develop an appreciation of the colour and design of natural habitats, we soon realize that Nature is an apt artist, not easily surpassed. True nature photographs are stunning and certainly possess the extra edge of honesty.

Above: *Danja on the point of releasing a porcupine pufferfish* (Diodon *sp.*), *just rescued from some novice divers who were unaware of its distress.*

Right: *Never break an animal's trust. Let marine life approach you when it is ready.*

DIVING'S BASIC CONSERVATION RULE

The basic rule for all divers is simply to leave things as they find them. This even means returning dead coral slabs to their original position after examination. The cover of these apparently dead slabs and rocks is vital in protecting all the thousands of eggs of fish, molluscs, crustaceans, and echinoderms. When the shelters are upturned and laid open, predation and tidal currents will destroy all promise of life. If the significance of this one simple act can be realized by all visitors to the reef – this includes divers, snorkellers, tide-pool waders, neap-tide strollers, and especially shell collectors – there will be a continued future for its myriad inhabitants.

PROMOTING ECOFRIENDLY DIVING

Our collective adherence to good practices on the reef must send a resounding message: we will not tolerate malice on our reefs. When we act together, we build a formidable rolling wave. But we must remember that this applies to bad attitudes too. If we tolerate abusive divers, we become part of the problem and our reputation is collectively tainted. Where such problems exist, we must address them ourselves. By policing ourselves, we forestall both guilt and outside regulation.

While stern finger-pointing might momentarily chastise a diver underwater, the obvious humiliation often leaves an unpleasant aftertaste and may lead to vengeful acts. There are many subtle and kind ways to reform the bad or ignorant guys, but strong action might be required for the persistent nasties. The diving community must have clear conservation policies and everyone should be made aware that bad behaviour will not be tolerated. Wilfully destructive divers, knife-pokers, animal harassers and graffiti artists deserve no sympathy and we certainly do not want them on the reefs.

They can be blacklisted and banned from air fills and dive boats. Proof of their bad behaviour can be shown on photographs or video at dive clubs and dive resorts, or even published in local newspapers and dive magazines. Any desire by such a diver to return to grace must be dependent on re-education and personal involvement in conservation efforts.

Dive schools should play a much stronger role. Diving is taught in short, intensive courses, largely governed by lack of time and commercial pressures. While the aim is to get divers diving, far too little time is spent on in-water body control and ecologically correct diving. This cannot be solved by the subsequent courses that are available from such schools. These are rarely subscribed to, as the extra expense seems unjustified to young, cash-strapped divers. It makes sense that correct behaviour be taught and tested as part of diver qualification. Schools that comply should be actively supported and promoted by all of us, while those who do not must be pressurized.

In the meantime there is only one way to get rid of clumsy novice divers on the reef. We must arrange social diving events at which we help teach buoyancy skills and correct reef behaviour. By drawing the beginners into our seasoned circles, we can create powerful communities with positive attitudes. It is rewarding, makes ecological sense, shows quick results and is far more fulfilling and effective than hysterical chastisement and criticism.

We should try to be ambassadors rather than just tourists when we dive in remote or unsophisticated areas. Very often the people at our dream destinations have not had the opportunity for education in environmental issues. Conservation may be unimportant to them in view of poverty and a struggle for survival. Active communication is the ideal way to learn more about the host country's people and culture, and to share your own in a hopefully positive way.

A good story is always welcomed by such people for its novelty and information. In places where we spend limited time but require maximum value, storytelling is perhaps the speediest short cut to obtaining the trust of local inhabitants. Our stories allow them to evaluate us and our intentions subtly. Amazing hospitality is frequently extended, based solely on judgements made during such times. A knowledge of local fish names and diet obviously benefits us as does prior advice about dangerous local species and injury treatment.

Simply say 'Did you know that…?' Share your knowledge and love of the sea in a noncondescending way – remember that illiteracy does not equate to a lack of intelligence. Do not raise your voice because you cannot speak their language. To preserve his dignity, the uneducated man may be adept at hiding his respect for people of learning, but believe me, he will store every one of your words in his memory and weigh each one against his own knowledge. With the goodwill that we create, we open doors to learning not only for them but also for ourselves. And we will certainly not only be judged by our words: thoughtless acts and disrespect to local customs may make us unwelcome. Leave enriched, having enriched.

Above: *Every diver should try to preserve the reefs intact for future generations to enjoy.*

As divers we see wondrous things, harbour intriguing secrets and experience a deep lifelong fulfilment that most landlubbers could never begin to understand. Let our tribute to the reefs be to preserve them. We are on a winning streak if every diver is able to control his own body and removes just one piece of litter on every dive. We can even practise successful repair work on the reef! I have seen countless examples of coral self-repair when divers take the time to right, prop up and support corals which were accidentally broken or knocked over in a storm.

Let us now become a vital and formidable voice for environmental consciousness and let us demonstrate it in all our actions, in water and on land.

FRIENDLY OCTOPUS AND SENSUAL NUDIBRANCHS

We have an excellent demonstration of how one simple story impacted on a diving community, rippled out and changed perceptions forever. Years ago, local dive leaders on a small island adopted a showy and rather brutal-looking method of amusing visiting divers. Day after day they extracted octopus from their burrows by inserting and wriggling a knife in the holes. Although the poor creatures were never harmed, they were left no choice but to pour out in alarm and be grabbed. Spurting copious clouds of ink, each octopus was wrestled into submission and then offered to the divers for patting and photography. Even so, the octopus were always returned safely to their homes, as the reef housed many eels.

I am convinced that these divemasters simply did not know better. Among the locals, recreational diving was still relatively exotic; divers were considered 'macho' and the 'octopus circus' was a natural extension of this perception. We knew that the practice would continue regardless, so a change of approach was all we could hope for. Knowing that criticism would be ineffective and alienating, we decided to befriend the leading diver, pointedly admiring his dive skills. We then jokingly told him that Danja could entice an octopus out of its house with her bare hands, and that it would voluntarily stay with her for part of a dive. The subtle manipulation worked! With much laughter and some back-slapping, a friendly challenge inevitably ensued.

I have to confess that Danja is an absolute wizard at getting creatures to cooperate with her willingly, but I quietly wondered if we were perhaps overoptimistic. She secretly tucked a tiny fish snack into her pocket on the next dive and searched for a less traumatized octopus. The gods of the reef must have wanted her to succeed. Soon, after a bit of patient 'hand-holding' between octopus and ecodiver, she managed to gently entice the creature into her cupped hands and then pursued some of her usual close eye-contact and telepathy to reassure it. Finally she plopped it onto her shoulder, free to hold on or flee. Miraculously, it chose to stay. It happily explored her ears and face with a dainty tentacle, accepted the proffered snack, and then climbed up on her head. For quite a while it tested the glass surface of her mask. Finally Danja dipped her head close to its burrow entrance and the octopus oozed inside. Everybody was enchanted. The divemaster was mesmerized. As soon as we surfaced he initiated a discussion. Then came the anticipated request: 'Teach me how to do that'. The 'magical' voluntary interaction appeared infinitely more impressive and to him, acquiring this talent was simply irresistible.

He buddied with Danja on all further dives and soon learnt the charm of gentle marine friendships. Many subsequent chats with him and his colleagues touched on the intricacies of the various creatures, hermaphroditism, male-to-female sex changes, spawning and commensalism. We explained why nudibranchs should not be removed from their chosen locality. Shrewdly, we put particular stress on the difficulty that these beautiful creatures have in locating sexual partners once removed from their home turf. To the sensual islanders, forced celibacy was a horrifying and unacceptable concept. Now, many years later, they diligently teach conservation in local schools. On the reefs they perpetuate the gentle interaction first learnt from us. They explain diver impact with great charm to their guests and are protective of their waters. Their reefs are in great condition without the necessity of regulation and the story of sexual abstinence for nudibranchs that are moved still regularly gets told!

Index

Page numbers in **bold** refer to photographs.

anemone 13, 30, 62, 89–97
bulb-tentacled (*Entacmaea quadricolor*) **90**
colonial (*Nemanthus annamensis*) **90**
tube (*Cerianthidae*) 57, 77, **91**
anemonefish *see* clownfish
angelfish (*Pomacanthidae*) 17, 33, 55
emperor (*Pomacanthus imperator*) **26**
anglerfish (*Antennarius nummifer*) **65**
anthias (*Serranidae*) 18, 29
argonaut (*Argonautidae*) 108

barracuda (*Sphyraenidae*) 33
barramundi cod (*Cromileptes altivelis*) 55
basal tendrils 86
bivalves 58–60
blennie (*Blenniidae*) 22, 54
midas (*Ecsenius midas*) 22
sabretooth (*Plagiotremus tapeinosoma*) 28, 32, **33**, **46**, 54, 101
body language 136
boxfish (*Ostraciidae*) 31, 55
branchiae 116
butterflyfish (*Chaetodontidae*) 17
masked (*Chaetodon semilarvatus*) **39**
redfin (*Chaetodon bifasciatus*) **12**

camouflage 25, 27, 144
cardinalfish (*Pogonidae*) 18, 38, 55, 77
goldstripe (*Apogon cyanosoma*) 18
catfish, eeltailed (*Plotosus lineatus*) **66**
cephalopods 103–111
chlorophyll 13
chromatophores 106
cirri 86
clams (*Tridacnidae*) 59, 60
cleaning stations 98–102
clingfish (*Discotrema crinophila*) 31, 88
clownfish (*Pomacentridae*) **80**, 89–97, 140
classification of 94
false (*Amphiprion ocellaris*) **37**, **95**
Mauritian (*Amphiprion chrysogaster*) **91**
panda (*Amphiprion polymnus*) 74, **97**
reproduction 95
saddleback *see* panda
spinecheek (*Premnas biaculeatus*) **95**
twoband (*Amphiprion bicinctus*) **89**
whitebonnet (*Amphiprion leucokranos*) **97**
Cnidaria 90
cnidoblasts 62
colour, importance of 25–26, 38
column 89
commensalism 41, 43
facultative 43
obligative 43
conservation 150–157
coral 30, 62
black (*Antipatharia* sp.) 49
bubble (*Plerogyra* sp.) **54**, 55
fire *see* hydroids
hard, *Diploastrea heliopora* **13**
mushroom (*Fungia* sp.) **13**, 55
soft (*Sarcophyton* sp.) **14**, 46–47, 52
(*Dendronephthya* sp.) 46, 47
stony 54–55
sunloving (*Euphyllia glabrescens*) 55
zooxanthius 65
coral cod (*Cephalopholis miniata*) **134**
corallimorpharians (*Discomatidae*) 62, 63
cornetfish, flutemouthed (*Fistularia commersonii*) 29
cowfish, longhorned (*Lactoria fornasini*) 74 (*see also* boxfish)
cowry (*Cypraeidae*) 47
allied (*Phenacovolva rosa*) **46**, 48
carnelian (*Cypraea caneola*) **58**
eggshell (*Ovula ovum*) 59
money (*Cypraea moneta*) **58**
toenail (*Calpurnus verrucosus*) 47
crab (*see also* crustaceans)
anemone hermit (*Dardanus pedunculatus*) **63**, 122
arrowhead (*Huenia heraldica*) 46
black coral (*Quadrella granulosa*) 49
(*Xenocarcinus conicus*) 49
boxer (*Lybia tessellata*) 63
decorator 46, 57, 71, 121
(*Camposcia retusa*) **121**
(*Naxoides taurus*) **46**, **121**
Dendronephthya (*Holophrys oatessi*) 46
depressed 122
(*Enicarcinus depressus*) 49
(*Xenocarcinus* sp.) **48**
hermit (*Diogenidae*) 63
(*Dardanus* sp.) **36**, **122**
coral (*Paguritta* sp.) **54**, 55
striped (*Trizopagurus strigatus*) 122
pea (*Xanthasia murigera*) 60
porcelain (*Porcellanidae* sp.) **97**
reef (*Carpilius convexus*) 77, **119**
sea star (*Lissecarcinus polyboides*) 120
spider (*Cyclocoeloma tuberculata*) **120**
sponge (*Dromiidae*) 122
spotted porcelain (*Neopetrolisthes* sp.) 97
urchin (*Zebrida adamsii*) 56
crinoids (*Comasteridae*) 73, 86–88
crocodilefish (*Cymbacephalus beauforti*) **38**, 70
crustaceans (*Crustacea*) 38, 46, 48, 54, 119–123
eggs 24
regenerative process 119
reproduction 120
currents 145
cuttlefish (*Sepiidae*) 29, 106–107, **112**
flamboyant (*Metasepia pfefferi*) **106**
mimicry 106

dactyls 57, 122
damselfish (*Pomacentridae*) 16, 55, 76
decapods 119
defence mechanisms 30–33
chemicals 30
darts 30
needles 30
poison 30
spines 30
depth indications, importance of 144
dive plan 37
diving safety 131
dolphin 132, **134**, **141**
bottlenose (*Tursiops truncatus*) **24**
dominofish (genus *Ascyllus*) 96

echinoderms 112
eel (*Muranidae*) 33, 77, 104
blue ribbon (*Rhinomuraena quaesita*) 74, 75
garden (*Heterocongridae* sp.) 66
moray (*Gymnothorax javanicus*) **100**, 101, 127, **129**
electric ray (*Narcinidae* sp.) 33, 70
epitoke 118
establishing trust 135

feather star *see* crinoid
feeding, ethics of 153
fertilization
cross 24
external 23
internal 24
self 23
filefish (*Pseudomonacanthus* sp.) 55, **86**
fish
marine 15
mouth-brooding 18
sex change 16, 18, 19, 96
spawning 16
territories 15
'fish trails' 134
flounder, peacock (*Bothus mancus*) **26**, **70**
frogfish *see* anglerfish
gladius 107

glassy sweepers (*Parapriacanthus guentheri*) **19**
goatfish (*Mullidae*) 70
gobies (*Gobiidae*) 22, 31, 48, 52
coral (*Pleurosicya mosambica*) 47
fire (*Nemateleotris magnifica*) 22
sanddarter (*Thrichonotis setiger*) 66
translucent (*Gobiidae* sp.) **49**
Trimmaton namis 22
wire coral (*Bryaninops yongei*) **53**
goldies *see* anthias
grouper (*Serranidae*) 18, 101
coral (*Cephalopholis* sp.) **98**
Queensland (*Epinephelus lanceolatus*) 18

hawkfish (*Cirrhitidae*)
Forster's (*Paracirrhites forsteri*) **39**
longnose (*Oxycirrhitus typus*) 48, **49**, 144
spotted (*Cirrhitichthys aprinus*) **15**
hermaphrodites 24, 60, 82
humming, effect of 136
hydroids (*Plumulariidae*) 30, 52, 53

jellyfish *see* sea jellies

leaf fish (*Scorpaenidae*) 31
sailfin (*Taenianotus triacanthus*) 27
Linnaeus 39
lionfish (*Pterois volitans*) 31, 77, 127, 138
dwarf (*Dendrochirus brachypterus*) **72**, 77
lobsters (*Palinuridae*)
elegant squat (*Allogalathea elegans*) **27**, **88**
reef (*Palinurella wieneckii*) 57, 77
rock (*Panulirus versicolor*) 57
slipper (*Parribacus caledonicus*) 55, 77

manta ray (*Manta birostris*) **102**
marine intelligence 126–130
marine relationships 132–149
mermaid's comb (*Murex pecten*) 66
mollusc (*Ophistobranchs*) 71, 77, 85
Moorish idol (*Zanclus cornutus*) 17

nautilus (*Nautilus pompilius*) **108**, **109**, **133**
paper *see* argonaut
nematocyst 30, 62
nudibranch (*Ophistobranchs*) 31, 47, 71, 77, 80, 156
harlequin (*Doridaceans*) 84
pyjama (*Chromodoris elizabethina*) **81**
side-gilled (*Dendrotaceans*) 84
Spanish dancer (*Hexabranchus sanguineus*) 80, **82**
tailgater (*Chromodoris* sp.) **81**
translucent (*Gymnodoris ceylonica*) **80**
tubercular (*Aeolidaceans*) 84
veiled (*Arminaceans*) 84

observation techniques 38
octopus (*Octopodidae*) 33, 77, 103–105, **140**, 156
blueringed (*Hapalochlaena* sp.) 33, 104
reproduction 104
operculum 116
oral disc 89
ovulid, Anga's (*Phenacovolva angasi*) 49
oysters, thorny (*Spondilidae*) 60

papillae 85
parrotfish (*Scaridae*) 19, **26**, 76, **77**
patience, importance of 137–140
pearlfish, translucent (*Fierasfer*) 61
pedal disc 89
pedicellariae 32
photography 133, 134, 137, 138, 142, 146-147
'peak of the shot' 139
photosynthesis 13
phytoplankton 14
pinnules 86
pipefish (*Syngnathinae*) 55
double-ended (*Syngnathoides biaculeatus*) 72, **73**
flagtailed (*Doryrhamphus multiannulatus*) 56
hairy ghost (*Solenostomus* sp.) **74**
ornate harlequin ghost (*Solenostomus paradoxus*) **29**, **39**, **43**, 73
seagrass ghost (*Solenostomus cyanopterus*) 73
plankton 13, 14, 48, 77
pleurobranchs (*Aplipsia* sp.) 82, **85**
polyp 46, 48, 54, **76**, 77, 121
predators 38
pufferfish (*Tetraodontidae*) 31, 55
masked (*Arothron nigropunctatus*) **32**, **101**
porcupine (*Diodon* sp.) 33, **153**

radioles 116
radula 81, 103
reproduction 23–24
rhinophores 80

sandperch, speckled (*Parapercis hexopthalma*) **66**
scientific classification 39
scorpionfish (*Scorpaenidae*)
decoy (*Iracundus signifer*) 28
Merlet's (*Rhinopias aphanes*) 27, **31**
tassled (*Scorpaenopsis oxycephalus*) **28**
sea cucumber 31, 60–61
feathermouth (*Synapta maculata*) **61**
Holothuria fuscopunctata **23**
leopard (*Bohadschia argus*) 61
peppermint-striped (*Thelenota rubrolineata*) 60
pineapple (*Thelenota ananas*) 60, 61
spiny black (*Sticopus chloronotus*) 61
sea hares *see* pleurobranchs
sea horse (*Hippocampinae*) 24, 72
Hippocampus kuda **39**
sea jellies (*Cnidaria*) **23**, 30, 62
'Belts of Venus' (*Scyphoza* sp.) 65
box (*Chironex fleckeri*) 65
Cubomedusae 65
mastigias (*Mastigias* sp.) **23**, **65**
upside-down (*Cassiopea* sp.) 61
sea moth, Pegasus (*Eurypegasus draconis*) 70
sea pens (*Pteroeidae*) **47**, 71
sea star (*Ophidiasteridiae*) 71, 112–115
basket 77, 114
blue (*Linckia laevigata*) **112**
brittle (*Ophioroidae*) 114
crown-of-thorns (*Acanthaster planci*) 32, 112, 113
feeding 112
Nardoa novacaledoniae **41**
reproduction 114
sea whips (*Elliselidae*) 52
sessile organisms 23, 38, 134
sharks (*Chondrichthyes*) 33
feeding 142, **143**
grey reef (*Charcharhinus amblyrhynchos*) **126**, 143
shell
bubble (*Hydatnidae* sp.) **58**
cone (*Coniidae*) **30**, 57
geographic (*Conus geographus*) 57
flamingo tongue (*Cyphoma gibbosum*) 47
golden wentletrap (*Epitonium billeeanum*) 47
harp (*Harpa* sp.) 71
helmet (*Cassis cornutus*) 58
spider (*Lambis* sp.) 58
triton (*Charonia tritonis*) 112
tun (*Tonna perdix*) 58
volute (*Volutidae*) 71
shrimp (*see also* crustaceans)
Amboin (*Thor amboinensis*) 62, 63
banded barber (*Stenopus hispidus*) 98
black coral (*Periclimenes psamathe*) 49
bubble cord (*Vir philippinensis*) **54**
cleaner (*Lysmata amboinensis*) **100**, **101**
Coleman's (*Periclimenes colmani*) 120
commensal (*Periclimenes soror*) **41**, 112, 113

corallimorpharian (*Pliopontonia furtiva*) **63**
crinoid (*Periclimenes* sp.) 88
eggshell anemone (*Periclimenes brevicarpalis*) **62**
glassy (*Periclimenes holthuisi*) **47**, **122**
gorgonian (*Tozeuma armatum*) 49
harlequin (*Hymenocera picta*) 57, **112**, 114
hingebeak (*Rhynchocinetes* sp.) **47**, 56, 120
imperial (*Periclimenes imperator*) **61**, 80, **82**
javelin mantis (*Harpiosquilla harpax*) 71, 74
javelin mantis (*Squillidae* sp.) **74**
mantis (*Odontodactylus scyllarus*) 33, 57, 122
needle (*Stigopontonia commensalis*) 56
painted cleaner (*Lysmata amboinensis*) **98**
skeleton (*Caprella* sp.) 122
snapping (*Sinalpheus stimpsoni*) 33, 88
striped lady (*Lysmata amboinensis*) 98
wire coral (*Dasycaris zanzibarica*) **28**, 52
Zanzibar *see* wire coral
shrimpfish (*Aeoliscus strigatus*) 55, **70**, 74
skills
scuba 130, 131, 151
snorkelling 133
snake star (*Astrobrachion adherens*) 49
sponges (*Spongiidae*) 30, 64–65
squid (*Sepiidae*) 107–108
squirrelfish (*Sargocentron spiniferum*) 32, **33**, 38
stingray (*Dasyatididae*) 32, 70
bluespotted (*Taeniura lymna*) **33**
stonefish (*Synanceia verrucosa*) 31
surgeonfish (*Acanthuridae*) 19
Acanthurus sp. **102**
sweetlips, spotted (*Plectorhinchus chaetodontoides*) **38**, **98**
symbiosis 41

touching, ethics of 150, 152
triggerfish (*Balistidae*) 33, 138
clown (*Balistoides conspicillum*) **139**
giant (*Balistoides viridescens*) **139**
Tropic of Cancer 8
Tropic of Capricorn 8
trout, coral (*Cephalopholis miniata*) 38, **137**
trumpetfish (*Aulostomus chinensis*) 29
tunicates 52, 53
turtles 137
underwater tools 40
unicornfish *see* surgeonfish 19
urchins (*Echinodermata*) 56, 72
fire (*Asthenosoma*) **32**, 72, 121
flower (*Toxopneustes pileolus*) 32

videography 146–147

waspfish (*Scorpaenidae*) 31
cockatoo (*Ablabys taenianotus*) 71
worms 54, 116–118
acorn (phylum *Hemichordata*) 66, 118
annelid polychaete 116
bottlebrush *see* Christmas-tree
bristle polychaete 77, 117
Christmas-tree (*Spirobranchus giganteus*) **116**, 134
feather (*Protula magnifica*) **117**
featherduster (*Sabellastarte magnifica*) 116, 134
flat (phylum *Platyhelminthes*) 118
palolo (*Eunice* sp.) 118
spaghetti (*Eupolimnia nebulosa*) 118
tube (*Phorona*) 116
wrasse (*Labridae*) 16
cleaner (*Labroides bicolor*) 101
cleaner (*Labroides dimidiatus*) 28, **102**
Thallasoma 97

yellowtail snapper (*Ocyurus chrysurus*) **126**

zooxanthellae 13, 61, 81, 82, 89, 90

PHOTOGRAPHIC CREDITS

All photographs © **Danja Köhler** with the exception of the following:
Mike Bacon: p153; **Juan Espi** (SIL): p40 (top left, centre left, bottom left, 2nd from bottom right, bottom right); **Brenton Geach** (*The Argus*): back cover (flap); **Anthony Johnson**: p40 (top right, 2nd from top right), 131 (bottom left, top right, 2nd from top right, centre right, 2nd from bottom right, bottom right); **Annemarie Köhler**: back cover (main photograph), pp73 (top right); **Flip Schulke** (Photo Access): p141; **Nancy Sefton** (Photo Access): p154; **Geoff Spiby**: pp130, 135, 136, 151 (top).

SIL = Struik Image Library